I0036292

COUNTING MOLES

SIMPLE SOLUTIONS - CHEMISTRY COUNTING MOLES

Edited By

Nigel P. Freestone

The University of Northampton
UK

BENTHAM SCIENCE PUBLISHERS LTD.
End User License Agreement (for non-institutional, personal use)

This is an agreement between you and Bentham Science Publishers Ltd. Please read this License Agreement carefully before using the ebook/echapter/ejournal (**"Work"**). Your use of the Work constitutes your agreement to the terms and conditions set forth in this License Agreement. If you do not agree to these terms and conditions then you should not use the Work.

Bentham Science Publishers agrees to grant you a non-exclusive, non-transferable limited license to use the Work subject to and in accordance with the following terms and conditions. This License Agreement is for non-library, personal use only. For a library / institutional / multi user license in respect of the Work, please contact: permission@benthamscience.org.

Usage Rules:

1. All rights reserved: The Work is the subject of copyright and Bentham Science Publishers either owns the Work (and the copyright in it) or is licensed to distribute the Work. You shall not copy, reproduce, modify, remove, delete, augment, add to, publish, transmit, sell, resell, create derivative works from, or in any way exploit the Work or make the Work available for others to do any of the same, in any form or by any means, in whole or in part, in each case without the prior written permission of Bentham Science Publishers, unless stated otherwise in this License Agreement.

2. You may download a copy of the Work on one occasion to one personal computer (including tablet, laptop, desktop, or other such devices). You may make one back-up copy of the Work to avoid losing it. The following DRM (Digital Rights Management) policy may also be applicable to the Work at Bentham Science Publishers' election, acting in its sole discretion:

- 25 'copy' commands can be executed every 7 days in respect of the Work. The text selected for copying cannot extend to more than a single page. Each time a text 'copy' command is executed, irrespective of whether the text selection is made from within one page or from separate pages, it will be considered as a separate / individual 'copy' command.
- 25 pages only from the Work can be printed every 7 days.

3. The unauthorised use or distribution of copyrighted or other proprietary content is illegal and could subject you to liability for substantial money damages. You will be liable for any damage resulting from your misuse of the Work or any violation of this License Agreement, including any infringement by you of copyrights or proprietary rights.

Disclaimer:

Bentham Science Publishers does not guarantee that the information in the Work is error-free, or warrant that it will meet your requirements or that access to the Work will be uninterrupted or error-free. The Work is provided "as is" without warranty of any kind, either express or implied or statutory, including, without limitation, implied warranties of merchantability and fitness for a particular purpose. The entire risk as to the results and performance of the Work is assumed by you. No responsibility is assumed by Bentham Science Publishers, its staff, editors and/or authors for any injury and/or damage to persons or property as a matter of products liability, negligence or otherwise, or from any use or operation of any methods, products instruction, advertisements or ideas contained in the Work.

Limitation of Liability:

In no event will Bentham Science Publishers, its staff, editors and/or authors, be liable for any damages, including, without limitation, special, incidental and/or consequential damages and/or damages for lost data and/or profits arising out of (whether directly or indirectly) the use or inability to use the Work. The entire liability of Bentham Science Publishers shall be limited to the amount actually paid by you for the Work.

General:

1. Any dispute or claim arising out of or in connection with this License Agreement or the Work (including non-contractual disputes or claims) will be governed by and construed in accordance with the laws of the U.A.E. as applied in the Emirate of Dubai. Each party agrees that the courts of the Emirate of Dubai shall have exclusive jurisdiction to settle any dispute or claim arising out of or in connection with this License Agreement or the Work (including non-contractual disputes or claims).
2. Your rights under this License Agreement will automatically terminate without notice and without the need for a court order if at any point you breach any terms of this License Agreement. In no event will any delay or failure by Bentham Science Publishers in enforcing your compliance with this License Agreement constitute a waiver of any of its rights.
3. You acknowledge that you have read this License Agreement, and agree to be bound by its terms and conditions. To the extent that any other terms and conditions presented on any website of Bentham Science Publishers conflict with, or are inconsistent with, the terms and conditions set out in this License Agreement, you acknowledge that the terms and conditions set out in this License Agreement shall prevail.

Bentham Science Publishers Ltd.
Executive Suite Y - 2
PO Box 7917, Saif Zone
Sharjah, U.A.E.
Email: subscriptions@benthamscience.org

**BENTHAM
SCIENCE**

CONTENTS

FOREWORD

When people I meet ask me what I do and I mention that I trained as a chemist, they usually wince. This is often followed by a comment such as, "there's too much maths in chemistry"; "learning chemistry is like learning a foreign language"; "there's a lot to learn". People very rarely say, "I wish I'd learned more chemistry at school" or "chemistry is so interesting". This is hugely frustrating for me as chemistry is everywhere – everything we see, touch, taste, smell and even hear involves chemicals and chemistry. I think it is a brilliant, fascinating, complex and simple subject that shapes and changes our world for the better.

However, having taught various aspects of (mainly environmental) chemistry from primary school to PhD levels for more than a quarter of a century, I'm very aware that chemistry can be viewed as difficult, theoretical and even irrelevant to people's everyday lives. This is a terrible shame and may have unforeseen adverse consequences. The world is facing a large number of global challenges that include anthropogenic climate change, sustainable development, the need for adequate supplies of clean water and other key resources such as metals, growing energy demands, concerns about food supply and quality, and the impacts on nutrition and health. Chemistry can make a significant contribution to helping us to mitigate or solve these challenges – but for that to happen, we need more chemists.

I know from talking to colleagues in industry, business and academia that there is a potential shortage of human resources in key scientific professions. There have been numerous pleas for the rejuvenation of science teaching and initiatives to increase the interest and attainment of students in science and mathematics. Basic scientific skills such as communication, classification, measurement, observation, inference and prediction contribute to a larger purpose - problem solving. These skills allow scientists to conduct systematic, objective investigations and to reach conclusions based on the results. Such skills are vital in chemistry. We need young people who are mathematically and scientifically literate and who engage with scientific problems in an orderly, "can-do" fashion. We cannot allow a shortfall in these skills to emerge simply through neglect or fatalism or worse – because subjects such as chemistry are "hard", "dangerous", "geeky", "too mathematical" or require "too much work".

Therefore I'm delighted to welcome Nigel Freestone's refreshing book "Counting Moles" to the stable of chemistry textbooks. Nigel has a wealth of experience in delivering chemical concepts to young people in an engaging and inspiring fashion. This book is designed to provide multiple opportunities for students to develop a firm understanding of chemical concepts that underpin the whole subject. Students will be able to use the multiple worked examples and calculating frame to hone their skills and become more confident in fundamental problem-solving activities, culminating in the award of a Counting Moles

Driving Licence that confirms their achievement.

I'm sure Nigel's book will make a significant contribution to addressing the chemistry skills shortage and to ensuring that this wonderful subject continues to make its distinct contribution to solving the world's challenges.

Ian Williams
University of Southampton
UK
I.D.Williams@soton.ac.uk

PREFACE

Students studying chemistry often struggle with the **MOLE.** This user-friendly self-teach package provides an effective aid to learning by giving clear and confident presentation of the essentials of the mole needed by those starting chemistry courses. This self teach package contains over 200 questions, with detailed solutions, so if you get stuck, you can see where you went wrong.

After successfully completing Counting Moles, you will be able to:

- understand what is meant by the terms relative atomic mass and relative formula mass
- calculate average atomic mass from isotope abundance;
- determine relative formula mass;
- define of the chemical term, the 'mole';
- calculate the molar mass of a substance from its chemical formula;
- calculate percentage composition;
- interconvert between mass, number of molecules, and moles;
- calculate the amount (mass or moles) of product expected to be formed in a chemical reaction, given the amounts of reactants used;
- calculate the amount (mass or moles) of reactants which need to be used in a chemical reaction in order to produce a specified amount of product;
- define the term concentration, and calculate the molarity of solutions from volumetric analysis and other data;
- determine the mass and/or volume of gases consumed or formed in a chemical reaction.

Counting Moles is split into SIX (6) Chapter. Each Chapter should be thought of as a separate lesson that should take between 15 and 45 minutes to complete. It is **ESSENTIAL** that you answer each question. The Counting Moles approach to learning is based on understanding developed through repetition and introduces the '**mole calculating frame**' to help you solve problems. Answers to ALL questions, with full explanations are provided, should you get into difficulty.

The Chapters prepare you for the **Counting Moles Driving Test**. If you PASS, you can throw you M-plate away and claim your **Counting Moles Driving Licence**. You will then be able to answer moles-based coursework and examination questions correctly and with confidence.

CONFLICT OF INTEREST

The author confirms that author has no conflict of interest to declare for this publication.

ACKNOWLEDGEMENTS

Declared none.

Copyright: Nigel P Freestone 2016

COUNTING MOLES

SIMPLE SOLUTIONS - CHEMISTRY COUNTING MOLES

Relative Atomic & Formula Mass, Percentage Composition, Empirical & Molecular Formulae

Keywords: Atoms, Average relative atomic mass, Calculations, Empirical and molecular formula, Isotopes, Percentage composition, Relative atomic mass (A_r), Relative formula mass (M_r).

1.1. RELATIVE ATOMIC MASS (A_R)

Chemical elements are the building blocks from which everything is constructed, from specks of dust to mobile phones and from flora and fauna to the clothes we wear. There are over a 100 known elements. An **element** is a pure substance that cannot be chemically broken down. The smallest unit of an element is the **atom**. Different atoms have different masses. The mass of an atom is so small that it is more convenient to compare atom masses, rather than refer to their actual mass. The standard for this relative scale is an atom of carbon-12, which has a relative atomic mass (A_r) of 12.

The Table below lists the relative atomic mass (A_r) values of some common elements.

Selected Relative Atomic Mass Values

Element	Approximate Relative Atomic Mass (A_r)
Hydrogen	1
Carbon	12
Nitrogen	14
Oxygen	16
Sodium	23
Magnesium	24
Silicon	28
Calcium	40

Nigel P. Freestone
All rights reserved-© 2016 Bentham Science Publishers

Table contd.....

Element	Approximate Relative Atomic Mass (A_r)
Bromine	80

These relative atomic mass values tell us for example that sodium atoms (A_r = 23) are 23 times heavier than hydrogen atoms (A_r = 1), two atoms of neon (A_r = 20) have the same mass as one atom of calcium (A_r = 40) and that three oxygen atoms (A_r = 16) weigh the same as two magnesium atoms (A_r = 24).

Relative atomic masses are listed in the periodic table.

Isotopes

Elements are defined by their proton (atomic) number. An atom with 7 protons is always nitrogen (N), an atom with 20 protons is always calcium and an atom with 70 protons must therefore always be gold (Au). Isotopes are atoms that have the same number of protons, but have different numbers of neutrons. For example, carbon has three naturally occurring isotopes, often referred to as simply carbon-12, carbon-13 and carbon-14, with relative atomic masses of 12, 13 and 14, respectively. Since carbon has a proton number of 6, the isotopes contain 6, 7 and 8 neutrons, respectively.

Three Isotopes of Carbon

Isotope	No. of protons	No of neutrons	No of electrons	Relative Atomic Mass (A_r)
$^{12}C_6$	6	6	6	12
$^{13}C_6$	6	7	6	13
$^{14}C_6$	6	8	6	14

A typical periodic table information box for the element carbon is given below:

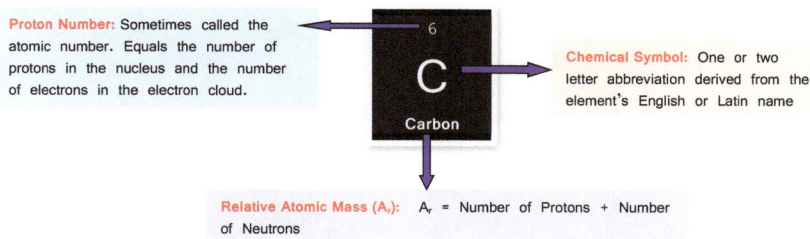

Proton Number: Sometimes called the atomic number. Equals the number of protons in the nucleus and the number of electrons in the electron cloud.

Chemical Symbol: One or two letter abbreviation derived from the element's English or Latin name

Relative Atomic Mass (A_r): A_r = Number of Protons + Number of Neutrons

The relative atomic masses listed in the periodic table are an average of the masses of the different naturally occurring isotopes. The rounded value gives you the mass of the most abundant isotope. For example, copper occurs naturally as Cu-63 and Cu-65. Given that the average relative atomic mass of copper is 63.546, we can conclude that Cu-63 is the more abundant isotope since the average value is closer to 63 than to 65.

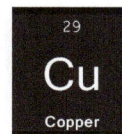

Calculating Average Relative Atomic Mass

Chlorine has a relative atomic mass of 35.5, which is the average of the masses of its two naturally occurring isotopes. This is calculated by working out the relative abundance of each isotope. For example, in any sample of chlorine, 25% will be Cl-37 and 75% Cl-35. The relative atomic mass is therefore calculated using the equation:

(% of isotope 1/100 × mass of isotope 1) + (% of isotope 2/100 × mass of isotope 2)

So in the case of chlorine:

Average relative atomic mass = (75/100 x 35) + (25/100 × 37) = 26.25 + 9.25 = **35.5**

Example 1.1: *Boron has three naturally occurring isotopes: 9% boron-10 and 80.1% boron-11. Calculate the relative atomic mass of boron.*

Answer:

Average relative atomic mass = (19.9/100 x 10) + (80.1/100 x 11) = 1.99 + 8.811 = **10.801**

Example 1.2: *Magnesium has three isotopes, Given that the natural abundances are Mg-24 (78.70%), Mg-25 (10.13%), and Mg-26 (11.17%) calculate the relative*

atomic mass of magnesium.

Answer:

Average relative atomic mass = $(78.7/100 \times 24) + (10.13/100 \times 25) + (11.17/100 \times 26)$

$$= 18.888 + 2.533 + 2.904$$
$$= \mathbf{24.33}$$

This problem can also be reversed, as in having to calculate the isotopic abundances when given the atomic mass and isotopic masses.

Example 1.3: *Nitrogen is made up of two isotopes, ^{14}N and ^{15}N. Given nitrogen's atomic weight of 14.007, what is the percentage abundance of each isotope?*

Answer:

Let x = percentage abundance of ^{14}N

Therefore, percentage abundance of ^{15}N = 100- x

$(x/100 \times 14) + (100-x/100 \times 15) = 14.007$

Multiply throughout by 100

$14x + 15(100 - x) = 1400.7$

$14x + 1500 - 15 x = 1400.7$

$x = 1500 - 1400.7 = 99.3$

Thus % abundance ^{14}N = 99.3%, ^{15}N = **0.7%**

Example 1.4: *The relative atomic mass of copper is 63.55. Copper has two naturally occurring isotopes, Cu-63 and Cu-65. Determine the natural abundance (%) of each isotope.*

Answer:

Let x = percentage abundance of ^{63}Cu

Therefore, percentage abundance of $^{65}Cu = 100- x$

$(x/100 \times 63) + (100-x/100 \times 65) = 63.55$

Multiply throughout by 100

$63x + 65(100 - x) = 6355$

$63x + 6500 - 65 x = 6355$

$2x = 6500 - 6355$

$x = 72.5$

Thus % abundance **Cu-63 = 72.5% and Cu-65 = 27.5%.**

Exercise 1.1

a. *The natural abundance for gallium isotopes is ^{69}Ga 60.11% and ^{71}Ga 39.89%. Calculate the relative atomic mass of gallium.*
b. *Neon has three isotopes, ^{20}Ne, ^{21}Ne and ^{22}Ne. The percentage of each in order is 90.48%, 0.27% and 9.25%. Calculate the relative atomic mass of neon.*
c. *Element Y exists as two naturally occurring isotopes, Y-51 and Y-52. Given that Y-51 has a natural abundance of 18%, calculate the relative atomic mass of element Y.*
d. *Lithium has two naturally occurring isotopes, Li-6 and Li-7. Determine the percentage abundance of each isotope, given that lithium has a relative atomic mass of 6.94.*
e. *Bromine has a relative atomic mass of 79.90. There are two known isotopes of bromine, with relative atomic masses of 79 and 81. Determine the percentage abundance of each isotope.*

NOTE: for simplicity A_r values used in the remainder of this text, with the exception of Cl (35.5) and Cu (63.5) will be integers.

1.2. RELATIVE FORMULA MASS (M_R)

Elements rarely occur alone in nature. Instead they form compounds by combining together. Chemical compounds are substances formed by two or more elements, in a fixed and constant proportion, held together by chemical bonds. For example, water (H_2O) comprises hydrogen and oxygen atoms in the ratio 2:1. If this ratio is changed to 1:1 then a new compound formed, the disinfectant and bleach, hydrogen peroxide (H_2O_2). Compounds have a definite formula with its constituents combined in fixed proportions. There are over 8 million known chemical compounds. The smallest unit of a compound is the **molecule.** Compounds contain two or more elements chemically combined. Thus all compounds (*e.g.* CH_4, $NaCl$, $CuSO_4$) are molecules, but not all molecules are compounds (*e.g.* H_2, O_2, P_4, S_8)

To find the relative formula mass (M_r) of a compound, you just add together the relative atomic mass (A_r) values for all the atoms in its formula:

Step 1: Identify the number of atoms of each element present;

Step 2: Find the relative atomic mass (A_r) for each element present;

Step 3: Calculate the total mass of each element present in the chemical formula of the compound, *i.e.* Mass of element in compound = relative atomic mass x number of atoms present;

Step 4: Add together all of the masses to obtain the Relative Formula Mass (M_r).

These steps are summarised in the following relative formula mass calculating frame for the compound, $X_aY_bZ_c$

Relative Formula Mass (M_r) Calculating Frame

Element	No. of atoms	A_r	Mass
X	a	$A_r(X)$	a x $A_r(X)$
Y	b	$A_r(Y)$	b x $A_r(Y)$
Z	c	$A_r(Z)$	c x $A_r(Z)$
Total			M_r

Example 1.5: *Determine the relative formula mass of sodium chloride (NaCl)?*

Step 1: Determine the number of atoms of each element present

Element	No. of atoms	A_r	Mass
Na	1		
Cl	1		
Total			

Step 2: Find the A_r for each element present

Element	No. of atoms	A_r	Mass
Na	1	23	
Cl	1	35.5	
Total			

Step 3: Calculate the total mass of each element present in the chemical formula of the compound

Element	No. of atoms	A_r	Mass
Na	1	23	1 x 23 = 23
Cl	31	35.5	1 x 35.5 = 35.5
Total			

Step 4: Add together all of the masses to obtain the Relative Formula Mass (M_r)

Element	No. of atoms	A_r	Mass
Na	1	23	1 x 23 = 23
Cl	31	35.5	1 x 35.5 = 35.5
Total			58.5

M_r [NaCl] = **58.5**

Example 1.6: *What is the relative formula mass of water (H_2O)?*

Step 1: Determine the number of atoms of each element present

Element	No. of atoms	A_r	Mass
H	2		
O	1		
Total			

Step 2: Find the A_r for each element present

Element	No. of atoms	A_r	Mass
H	2	1	
O	1	2	
Total			

Step 3: Calculate the total mass of each element present in the chemical formula of the compound

Element	No. of atoms	A_r	Mass
H	2	1	2 x 1 = 2
O	1	16	1 X 16 = 16
Total			

Step 4: Add together all of the masses to obtain the Relative Formula Mass (M_r)

Element	No. of atoms	A_r	Mass
H	2	1	2 x 1 = 2
O	1	16	1 X 16 = 16
Total			18

Mr [H_2O] = **18**

Example 1.7: *Calculate the relative formula mass (M_r) of Al_2O_3*

Answer:

Step1: Determine the number of atoms of each element present

Element	No. of atoms	A_r	Mass
Al	2		
O	3		
Total			

Step 2: Find the A_r for each element present

Element	No. of atoms	A_r	Mass
Al	2	27	
O	3	16	
Total			

Step 3: Calculate the total mass of each element present in the chemical formula of the compound

Element	No. of atoms	A_r	Mass
Al	2	27	2 x 27 = 54
O	3	16	3 x 16 = 48
Total			

Step 4: Add together all of the masses to obtain the Relative Formula Mass (M_r)

Element	No. of atoms	A_r	Mass
Al	2	27	2 x 27 = 54
O	3	16	3 x 16 = 48
Total			102

$M_r [Al_2O_3] = $ **102**

Example 1.8: *Determine the relative formula mass (M_r) of $Mg(OH)_2$*

Answer:

Element	No. of atoms	A_r	Mass
Mg	1	24	1 x 24 = 24
O	2	16	2 x 16 = 32
H	2	1	2 x 1 = 2
Total			58

Note: $Mg(OH)_2$ could be rewritten as MgO_2H_2

$M_r [Mg(OH)_2] = $ **58**

Example 1.9: *Determine the relative formula mass (M_r) of Ca(NO$_3$)$_2$*

Answer:

Element	No. of atoms	A$_r$	Mass
Ca	1	40	1 x 40 = 40
N	2	14	2 x 14 = 28
O	6	16	6 x 16 = 96
Total			164

Note: $Ca(NO_3)_2$ can be rewritten as CaN_2O_6
M_r [Ca(NO$_3$)$_2$] = **164**

Example 1.10: *Determine the relative formula mass (M_r) of ammonium sulphate, (NH$_4$)$_2$SO$_4$?*

Answer:

Element	No. of atoms	A$_r$	Mass
N	2	14	2 x 14 = 28
H	8	1	8 x 1 = 8
S	1	32	1 x 32 = 32
O	14	16	4 x 16 = 64
Total			132

M_r [(NH$_4$)$_2$SO$_4$] = **132**

Exercise 1.2

What is the relative formula mass (M_r) of the following chemicals?

a. *Carbon dioxide, CO$_2$*
b. *Iron (II) sulphide, FeS*
c. *Copper sulphate, CuSO$_4$*
d. *Benzene, C$_6$H$_6$*
e. *Calcium hydroxide, Ca(OH)$_2$*
f. *Oxygen, O$_2$*
g. *Sodium oxide, Na$_2$O*
h. *Lead (II) nitrate, Pb(NO$_3$)$_2$*
i. *Nitrogen dioxide, NO$_2$*

　　j. *Ethanoic acid, CH_3COOH*
　　k. *Aluminium sulphate, $Al_2(SO_4)_3$*
　　l. *Iron (III) nitrate, $Fe(NO_3)_3$*
　m. *Calcium phosphate, $Ca_2(PO_4)_3$*
　　n. *Silver nitrite, $AgNO_2$*
　　o. *Potassium permanganate, $KMnO_4$*
　　p. *Potassium dichromate, $K_2Cr_2O_7$*
　　q. *Nickel sulphite, $NiSO_3$*
　　r. *Copper tartrate, $Cu_2C_4H_4O_6$*
　　s. *Cobalt (II) chlorate, $Co(ClO_3)_2$*
　　t. *Diethyl zinc, $(C_2H_5)_2Zn$*

1.3 PERCENTAGE COMPOSITION (%)

The percentage composition of a chemical compound is the percentage by mass of each constituent element.

To calculate the percentage composition of a chemical compound:

　　　　Step 1: Determine the number of atoms of each element present;
　　　　Step 2: Find relative atomic mass (A_r) for each element present;
　　　　Step 3: Calculate the total mass of each element present in the chemical formula of the compound, *i.e.*
　　　　Mass of element in compound = relative atomic mass x number of atoms present;
　　　　Step 4: Add together all of the masses to obtain the Relative Formula Mass (M_r);
　　　　Step 5: For each element, calculate its percentage composition by mass by:

　　　　　　　Mass of element in the compound/M_r x 100
　　　　and check that the % composition of each element add up to 100.

If the values do not add up to 100 then you have made an error and will need to repeat the calculation.

These steps are summarised in the following *percentage composition calculating frame* for $X_aY_bZ_c$

Percentage Composition Calculating Frame

Answer:

Element	No. of atoms	A_r	Mass	% composition
X	a	$A_r(X)$	a x $A_r(X)$	a x $A_r(X)/M_r$ x 100
Y	b	$A_r(Y)$	b x $A_r(Y)$	b x $A_r(Y)/M_r$ x 100
Z	c	$A_r(Z)$	c x $A_r(Z)$	c x $Ar(Z)/M_r$ x 100
Total			M_r	100%

Example 1.11: *Calculate the percentage composition of* Al_2O_3

Answer:

Steps 1-4 are the same as for calculating the relative formula mass of a compound.

Element	No. of atoms	A_r	Mass	% composition
Al	2	27	54	
O	3	16	48	
Total			102	

Step 5:

Element	No. of atoms	A_r	Mass	% composition
Al	2	27	54	54/102 x 100 = 52.9
O	3	16	48	48/102 x 100 = 47.1
Total			102	100%

Al_2O_3 has the following composition: **Al 59.2% and O 47.1%**

Example 1.12: *Calculate the percentage composition of* Na_2CO_3

Answer:

Steps 1-4 are the same as for calculating the relative formula mass of a compound.

Element	No. of atoms	A_r	Mass	% composition
Na	2	23	46	
C	1	12	12	
O	3	16	48	
Total			106	

Step 5:

Element	No. of atoms	A_r	Mass	% composition
Na	2	23	46	46/106 x 100 = 43.4
C	1	12	12	12/106 x 100 = 11.3
O	3	16	48	48/106 x 100 = 45.3
Total			106	100%

Na_2CO_3 has the following composition: **Na 43.4%, C 11.3% and O 45.3%**

Example 1.13: *What is the mass of copper present in 150 tonnes of the ore, chalcopyrite $CuFeS_2$?*

Answer:

Element	No. of atoms	A_r	Mass	% composition
Cu	1	63.5	63.5	63.5/183.5 x 100 = 34.6
Fe	1	19	56	56/183.5 x 100 = 30.5
S	2	32	64	64/183.5 x 100 = 34.9
Total			183.5	100%

The composition of Cu in $CuFeS_2$ is 34.6%.
Thus 150 tonnes will contain 34.6/100 x 150 = **51.9 tonnes of copper.**

Example 1.14: *1.26 g of iron reacts with 0.54 g of oxygen to form rust. What is the percentage composition of each element in the new compound?*

Answer:

This type of question can be answered using a shortened version of the same calculating frame.

Mass of new compound = 1.26 + 0.54 = 1.8 g

Element	Mass	% composition
Fe	1.26	1.26 /1.8 x 100 = 70
O	0.54	0.54/1.8 x 100 = 30
Total	1.6	100%

Rust has the following composition: **Fe 70%, and O 30%.**

Example 1.15: *Decomposition 8.657 g of a liquid sample into its elements gave 5.217g of carbon , 0.9620 g of hydrogen and 2.678 g of oxygen. Determine the percentage composition of the liquid.*

Answer:

Element	Mass	% composition	
C	5.217	1.26 /1.8 x 100 = 70	
H	0.962	0.962/8.657 x 100 = 11.1	
O	2.478	2.478/8.657x 100 = 28.6	
Total		100%	

The compound has the following percentage composition: **C 60.3%, H 11.1% and O 28.6%**

Exercise 1.3

a. *What is the percentage composition by mass of silicon and chlorine in $SiCl_4$?*

b. *Calculate percentage composition of $CuSO_4$*

c. *Calculate the mass percentage of hydrogen in aspirin, $C_9H_8O_4$.*

d. *What is the mass of silicon in 10g of clay, $Al_2Si_2O_5(OH)_4$?*

e. *What is the mass of sulphur in 1 tonne of H_2SO_4?*

f. *Determine the percentage composition of $Ca_3(PO_4)_2$*

g. *What is the percentage composition of ammonium sulfate, $(NH_4)_2SO_4$?*

h. *What is the mass of nitrogen present in 5g of aniline, $C_6H_5NH_2$?*

i. *9.03g of Mg combine completely with 3.48g of N to form a compound. What is the percentage composition of this compound?*

j. *A 27.0 g sample of a compound contains 7.20 g of C, 2.20 g of hydrogen and 17.6 g of oxygen. Calculate the percentage composition of the compound.*

1.4. EMPIRICAL AND MOLECULAR FORMULA

The empirical formula gives the simplest integer ratio of each element present in a compound. The integers are given as subscripts to right hand side of the chemical symbols. A molecular formula is the same as or a multiple of the empirical formula, *i.e.* $(CH_2O)n$. Thus if a compound has an empirical formula of CH_2O its molecular formula could be CH_2O (n= 1), $C_2H_4O_2$ (n= 2), $C_4H_8O_2$ (n = 4) or $C_{10}H_{20}O_{10}$ (n = 10) *etc.*

Empirical Formula from Molecular Formula

In general the empirical formula is obtained by dividing the subscripts of a molecular formula by the highest common denominator. This is best illustrated by the following examples:

The molecular formula of glucose is $C_6H_{12}O_6$. The highest common denominator of the subscripts, 6, 12 and 6, is 6. If we divide the subscripts in the molecular formula of glucose by 6, we get its empirical formula, CH_2O.

Hydrogen peroxide has the molecular formula: H_2O_2. The highest common denominator of the subscripts 2 and 2 is 2. Thus the empirical formula of hydrogen peroxide is HO.

Water has the molecular formula, H_2O. Since the highest common denominator of the subscripts is 1, the empirical and molecular formulas for water are the same.

Exercise 1.4

Determine the empirical formula from the following molecular formula:

a. C_6H_6
b. $C_2H_4O_2$
c. P_4O_6
d. N_2O_5
e. C_6H_9
f. CH_2OHCH_2OH
g. $C_6H_8O_6$

h. CuC_2O_4

i. Hg_2F_2

j. N_2O_4

Molecular Formula from Empirical Formula

Remember that the molecular formula is simply a multiple of the empirical formula. Thus the relative molecular mass of the molecular formula will be a multiple of the relative molecular mass of the empirical formula *i.e.* $(X_aY_bZ_c)_n$

Example 1.16: *What is the molecular formula of a compound that has a molecular mass of 42 and an empirical formula* CH_2

Answer:

Relative molecular mass of $CH_2 = 12 + (2 \times 1) = 14$

The number of multiples of empirical formula (*i.e.* n) = M_r (Molecular Formula)/M_r (Empirical Formula) = $42/14 = 3$

Molecular formula = $(CH_2)_3 = \mathbf{C_3H_6}$

Example 1.17: *Determine the molecular formula vitamin C given that the empirical formula* $C_3H_4O_3$ *and has a relative molecular mass of 176.*

Answer:

Relative molecular mass of $C_3H_4O_3 = 88$

Molecular formula = $(C_3H_4O_3)_n$

$n = 176/88 = 2$

Molecular formula = $(C_3H_4O_3)_2 = \mathbf{C_6H_8O_6}$

Exercise 1.5

a. *What is the molecular formula of a substance which has the empirical formula* CH_2O *and a molecular mass of 180?*

b. *Determine the molecular formula of chemical compound with a molecular* mass of 46.0 and an empirical formula of NO_2.

c. Which chemical compound has the empirical formula PO_2 and a relative formula mass of 252?

d. What is the molecular formula of a compound with an empirical formula $C_2H_4O_4$ and a relative formula mass of 132?

e. A compound with an empirical formula of C_4H_4O and a relative formula mass of 272. What is the molecular formula of this compound?

f. Determine the molecular formula of a compound with the empirical formula of CFBrO and a relative formula mass of 254.7.

g. What is the molecular formula of a compound with an empirical formula of CH_2N and a molecular mass of 84?

h. Dimethylglyoxime, has the empirical formula C_2H_4NO and a relative formula mass is 116. What is the molecular formula of the compound?

i. The common headache remedy, Ibuprofen has the empirical formula C_7H_9O and a molecular mass of approximately 218. Determine the molecular formula of Ibuprofen.

j. Determine the molecular formula of a compound with an empirical formula of C_2H_3O and a molecular mass of 172.

Empirical and Molecular Formulas from % Composition

To calculate empirical formula from percentage compositions of a compound use the calculating frame given below:

Empirical Formula from Percentage Composition Calculating Frame

	Element 1	Element 2	Element 3
% composition			
A_r			
% composition/A_r			
Ratio			

Step 1: Write the names or symbols of the elements;
Step 2: For each element give its % composition;
Step 3: Find the A_r value for each element;
Step 4: Divide the % value for each element by its A_r;

Step 5: Divide throughout by the smallest value;

Step 6: Write down the chemical formula.

The action at Step 5 usually gives you the simplest whole number ratio straightaway. Sometimes it does not, so you might get 1 and 1.5. In this example, you would multiply both numbers by 2, giving 2 and 3 (instead of rounding 1.5 up to 2).

Example 1.8: *What is the empirical formula of the compound with the following composition (by mass): 48.38% carbon, 8.12% hydrogen and 43.5% oxygen?*

Answer:

$\%O = 100 - 48.38 - 8.12 = 43.5\%$

Step 1		**Carbon**	**Hydrogen**	**Oxygen**
Step 2	% composition	48.38	8.12	43.5
Step 3	A_r	12	1	16
Step 4	% composition/A_r	48.38/12 = 4.03	8.12/1 = 8.12	43.5/16 = 2.7
Step 5	Ratio	1.5	3	1
		3	6	2

Empirical Formula: $C_3H_6O_2$

Example 1.9: *Determine the empirical formula and the molecular formula of a compound that contains 50.05 % sulfur and 49.95 % oxygen by weight and has a molecular mass of 64.07.*

Answer:

Step 1		**Sulfur**	**Oxygen**
Step 2	% composition	50.05	49.95
Step 3	A_r	32	16
Step 4	% composition/A_r	50.05/32.06 = 1.156	49.95/16 =3.12
Step 5	Ratio	1.156	3.12
		1	2

Empirical Formula: SO_2

Example 1.20: *A compound with a molecular mass of 74.14 was found to contain 64.80 % carbon, 13.62 % hydrogen, and 21.58 % oxygen by weight. Determine the empirical formula and the molecular formula of this compound.*

Answer:

Step 1		Carbon	Hydrogen	Oxygen
Step 2	% composition	64.8	13.62	21.58
Step 3	A_r	12	1	16
Step 4	% composition/A_r	64.8/12 = 5.4	13.62/1= 13.62	21.58/16 = 1.34
Step 5	Ratio	4	10	1

Empirical Formula: $C_4H_{10}O$

Example 1.21: *An antimony chloride was formed by completely reacting 0.295 g of chlorine with a 0.338 g sample of antimony. What is the empirical formula of the antimony chloride?*

Answer:

Mass of antimony chloride = 0.338 + 0.295 = 0.633 g

Step 1		Antimony	Chlorine
Step 2	% composition	0.338/0.633 x 100 = 53.4	0.295/0.633 = 46.6
Step 3	A_r	121.8	35.5
Step 4	% composition/A_r	53.4/121.8 = 0.44	46.6/35.5 = 1.31
Step 5	Ratio	0.44 1	1.31 3

Empirical formula: $SbCl_3$

Example 1.22: *A sample of magnetite contained 50.4 g of iron and 19.2 g of oxygen. Calculate the empirical formula of the ore.*

Answer:

Mass of magnetite sample = 50.4 + 19.2 = 69.6 g
Mass of magnetite sample = 50.4 + 19.2 = 69.6 g
% Fe = 50.4/69.6 x 100 = 72.4%

%O = 19.2/69.6 x 100 = 27.6%

Step 1		Iron	Oxygen
Step 2	% composition	72.4	27.6
Step 3	A_r	55.85	35.5
Step 4	% composition A_r	1.3	1.73
Step 5	Ratio	1	1.33
		3	4

Empirical formula: Fe_3O_4

Exercise 1.6

a. *A compound contains 24.74% potassium, 34.76% manganese, and 40.50% oxygen. Determine the empirical formula of this compound.*

b. *The percent composition of a compound was found to be 19.3% sodium, 26.9% sulfur, and 53.8% oxygen. Its formula mass is 238. What is its molecular formula?*

c. *A compound contains 16.7 g of Iridium and 10.3 g of selenium, what is its empirical formula?*

d. *It is found that 207g of lead combined with 32g of sulphur to form 239g of lead sulphide. From the data work out the formula of lead sulphide.*

e. *Find the empirical formula of a compound which has the following percent composition 41.3% C, 10.4% H, and 48.2% N.*

f. *Determine the empirical and molecular formula for a compound if its molar mass is 308 and contains 85.69% carbon; 3.93% hydrogen and 10.38% oxygen.*

g. *Determine the empirical and molecular formulas of the female steroid hormone, estradiol, which has a molecular mass of 278.*

h. *A sample of the amino acid glycine contains 1.56 g C, 0.333 g H, 2.08 g O, and 0.910 g N. Determine its empirical formula.*

i. *The composition of acetic acid is found to be 39.9% C, 6.7% H and 53.4% O. Determine the empirical formula of acetic acid.*

j. *A 50.51 g sample of a phosphorus chloride was decomposed. Analysis of the products showed that 11.39 g of phosphorus was produced. What is the empirical formula of the chloride?*

k. *What is the empirical formula of a molecule containing 18.7% lithium, 16.3%*

carbon, and 65.0% oxygen?

l. *Use the following percent composition data to determine the empirical formula of an unknown compound: 40.50% Zn; 19.86% S; 39.64% O.*

m. *A hydrocarbon contains 40.10% carbon; 6.49% hydrogen and 51.41% oxygen. Determine the empirical formula of the compound.*

n. *A nitrogen oxide contains 36.84% N. What is the empirical formula of the oxide?*

o. *An unknown compound was found to have the following percent composition: 47.0 % potassium, 14.5 % carbon, and 38.5 % oxygen. What is its empirical formula? If the relative molecular mass of the compound is 166, what is its molecular formula?*

ANSWERS

Exercise 1.1

a. *The natural abundance for gallium isotopes is* ^{69}*Ga 60.11% and* ^{71}*Ga 39.89%. Calculate the atomic weight of gallium.*

Answer:

Average relative atomic mass = (60.11/100 x 69) + (39.89/100 x 71)
$$= 41.48 + 28.32 = 69.8$$

b. *Neon has three isotopes,* ^{20}*Ne,* ^{21}*Ne and* ^{22}*Ne. The percentage of each in order is 90.48%, 0.27% and 9.25%. Calculate the relative atomic mass of neon.*

Answer:

Average relative atomic mass = (90.48/100 x 20) + (0.27/100 x 21) + (9.25/100 x 22)
$$= 19.096 + 0.57 + 2.09 = 21.76$$

c. *Element Y exists as two naturally occurring isotopes, Y-51 and Y-52. Give that Y-51 has a natural abundance of 18%, calculate the relative atomic mass of element Y.*

Answer:

Average relative atomic mass = (18/100 x 51) + (82/100 x 52)
$$= 9.18 + 42.64$$
$$= 51.82$$

d. *Lithium has two naturally occurring isotopes, Li-6 and Li-7. Determine the percentage abundance of each isotope, given that lithium has a relative atomic mass of 6.94.*

Answer:

Let x = percentage abundance of Li-6
Therefore, percentage abundance of ^{7}Li = 100- x
Thus, (x/100 x 6) + (100-x/100 x 7) = 6.94
Multiply throughout by 100
6x + 7(100 - x) = 694
6x + 700 – 7x = 694
x = 700 – 694 = 7
Thus % abundance Li-6 = 7%, Li-7 = 93%

e. *Bromine has a relative atomic mass of 79.90. There are two known isotopes of bromine, with relative atomic masses of 79 and 81. Determine the percentage abundance of each isotope.*

Answer:

Let x = percentage abundance of Br-79
Therefore, percentage abundance of Br-81 = 100- x
Thus, (x/100 x 79) + (100-x/100 x 81) = 79.90
Multiply throughout by 100
79x + 81(100 - x) = 7990

$79x + 8100 - 81x = 7990$

$2x = 8100 - 7990 = 110$

$x = 55$

Thus % abundance Br-79 = 55%, Br-81 = 45%

Exercise 1.2

What is the relative formula mass (M_r) of the following chemicals?

a. CO_2

Answer:

Element	No. of atoms	A_r	Mass
C	1	12	12
O	2	16	32
Total			44

$M_r [CO_2]$ = **44**

b. *FeS*

Answer:

Element	No. of atoms	A_r	Mass
Fe	1	56	56
S	1	32	32
Total			88

$M_r [FeS]$ = **88**

c. $CuSO_4$

Answer:

Element	No. of atoms	A_r	Mass
Cu	1	63.5	63.5
S	1	32	32
O	4	16	64

Table contd.....

Element	No. of atoms	A_r	Mass
Total			159.5

M_r [CuSO$_4$] = **159.5**

d. C_6H_6

Answer:

Element	No. of atoms	A_r	Mass
C	6	12	72
H	6	1	6
Total			78

M_r [C$_6$H$_6$]= **78**

e. $Ca(OH)_2$

Answer:

Element	No. of atoms	A_r	Mass
Ca	1	40	40
O	2	16	32
H	2	1	2
Total			74

M_r [Ca(OH)$_2$] = **74**

f. O_2

Answer:

Element	No. of atoms	A_r	Mass
O	2	16	32
Total			32

M_r [O$_2$] = **32**

g. Na_2O

Answer:

Element	No. of atoms	A_r	Mass
Na	2	23	46
O	1	16	16
Total			62

M_r [Na_2O] = **62**

h. $Pb(NO_3)_2$

Answer:

Element	No. of atoms	A_r	Mass
Pb	1	12	12
N	2	14	28
O	6	16	96
Total			331

M_r [$Pb(NO_3)_2$] = **331**

i. NO_2

Answer:

Element	No. of atoms	A_r	Mass
N	1	14	14
O	2	32	32
Total			46

M_r [NO_2] = **46**

j. CH_3COOH

Answer:

Element	No. of atoms	A_r	Mass
C	2	12	24

Table contd.....

Element	No. of atoms	A_r	Mass
H	4	1	4
O	2	16	32
Total			60

M_r [CH_3COOH] = **60**

k. $Al_2(SO_4)_3$

Answer:

Element	No. of atoms	A_r	Mass
Al	2	27	54
S	3	32	96
O	12	16	192
Total			342

M_r [$Al_2(SO_4)_3$] = **342**

l. $Fe(NO_3)_3$

Answer:

Element	No. of atoms	A_r	Mass
Fe	1	56	56
N	3	14	42
O	9	16	144
Total			242

M_r [$Fe(NO_3)_3$] = **242**

m. $Ca_3(PO_4)_2$

Answer:

Element	No. of atoms	A_r	Mass
Ca	3	40	120
P	2	31	62
O	8	16	128
Total			310

M_r [$Ca_3(PO_4)_2$] = **310**

n. $AgNO_2$

Answer:

Element	No. of atoms	A_r	Mass
Ag	1	108	108
N	1	14	14
O	2	16	32
Total			331

M_r [$AgNO_2$] = **154**

o. $KMnO_4$

Answer:

Element	No. of atoms	A_r	Mass
K	1	39	39
Mn	1	55	55
O	4	16	64
Total			158

M_r [$KMnO_4$] = **158**

p. $K_2Cr_2O_7$

Answer:

Element	No. of atoms	A_r	Mass
K	2	39	78
Cr	2	52	104
O	7	16	112
Total			294

M_r [$K_2Cr_2O_7$] = **294**

q. $NiSO_3$

Answer:

Element	No. of atoms	A_r	Mass
Ni	1	59	59
S	1	32	32
O	3	16	48
Total			139

M_r [$NiSO_3$] = **139**

r. $Cu_2C_4H_4O_6$

Answer:

Element	No. of atoms	A_r	Mass
Cu	2	63.5	127
C	4	12	48
H	4	1	4
O	6	16	96
Total			275

M_r [$Cu_2C_4H_4O_6$] = **275**

s. $Co(ClO_3)_2$

Answer:

Element	No. of atoms	A_r	Mass
Co	1	59	59
Cl	2	35.5	71
O	6	16	96
Total			226

M_r [$Co(ClO_3)_2$] = **226**

t. $(C_2H_5)_2Zn$

Answer:

Element	No. of atoms	A_r	Mass
Zn	1	65	65
C	4	12	48
H	10	1	10
Total			331

$M_r [(C_2H_5)_2Zn] = $ **123**

Exercise 1.3

a. *What is the percentage composition by mass of silicon and chlorine in $SiCl_4$?*

Answer:

Element	No. of atoms	A_r	Mass	% composition
Si	1	28	28	$28/170 = 16.5$
Cl	4	35.5	142	$142/170 = 83.5$
Total			170	100%

$SiCl_4$ has the following composition: **Si 16.5% and Cl 83.5%**

b. *Calculate percentage composition of $CuSO_4$*

Answer:

Element	No. of atoms	A_r	Mass	% composition
Cu	1	63.5	63.5	$63.5/1 = 39.8$
S	1	32	32	$32./159.5 = 20.1$
O	4	16	64	$64/159.5 = 40.1$
Total			159.5	100%

$CuSO_4$ has the following composition: **Cu 39.8%, S 20.1%, and O 40.1%**

c. *Calculate the mass percentage of hydrogen in aspirin, $C_9H_8O_4$.*

Answer:

Element	No. of atoms	A_r	Mass	% composition
C	9	12	108	108/180 x 100 = 60
H	8	1	8	8/180 x 100 = 4.5
O	4	16	64	64/180 x 100 = 35.5
Total			180	100%

$C_9H_8O_4$ has the following percentage composition: **C 60.0%, H 4.5% and O 35.5%**

d. *What is the mass of silicon in 10g of clay, $Al_2Si_2O_5(OH)_4$?*

Answer:

Element	No. of atoms	A_r	Mass	% composition
Al	2	27	54	20.9
Si	2	28	56	21.7
O	9	16	144	55.8
H	4	1	4	1.6
			258	100%

*$Al_2Si_2O_5(OH)_4$ has the following composition: Al 20.9%, Si 21.7%, O 55.8% and H 1.6% Thus, 10 g of clay will contain 21.7/100 x 10 = **2.17 g***

e. *What is the mass of sulphur in 1 tonne of H_2SO_4?*

Answer:

Element	No. of atoms	A_r	Mass	% composition
H	2	1	2	2/98 x 100 = 2
S	1	3	32	32/98 x 100 = 32.7
O	4	16	64	64/98 x 100 = 65.3
Total			98.	100%

Thus the mass of S in 1 tonne of H_2SO_4 = 32.7/100 x 1 = 0.327 tonne = **327 kg**

f. *Determine the percentage composition of* $Ca_3(PO_4)_2$

Answer:

Element	No. of atoms	A_r	Mass	% composition
Ca	3	40	120	120/310 x 100 = 38.7
P	2	31	62	62/310 x 100 = 20
O	8	16	128	128/310 x 100 = 41.3
Total			310	100%

$Ca_3(PO_4)_2$ has the following composition: **Ca 38.7%, P 20.0% and O 41,3%**

g. *What is the percentage composition of ammonium sulfate,* $(NH_4)_2SO_4$*?*

Answer:

Element	No. of atoms	A_r	Mass	% composition
N	2	14	28	28/132x 100 = 21.2
H	8	1	8	8/132 x 100 = 6.1
S	1	32	32	32/132 x 100 = 24.2
O	4	16	64	64/132 x 100 = 48.5
Total			132	100%

$(NH_4)_2SO_4$ has the following composition: **N 21.2%, H 6.1%, S 24.2% and O 48.5%**

h. *What is the mass of nitrogen present in 5 g of aniline,* $C_6H_5NH_2$*?*

Answer:

Element	No. of atoms	A_r	Mass	% composition
C	6	12	72	72/93x 100 = 77.4
H	7	1	7	7/93 x 100 = 7.5
N	1	14	14	14/93 x 100 = 15.1
Total			93	100%

$C_6H_5NH_2$ has the following percentage composition: C 77.4%, H 7.5% and N 15.1%
Thus 1g of $C_6H_5NH_2$ contains 15.1/100 x 1 = 0.151 g of N
Therefore 5 g $C_6H_5NH_2$ contains 0.151 x 5 = **0.755 g of N**

i. *9.03 g of Mg combine completely with 3.48 g of N to form a compound. What is the percentage composition of this compound?*

Answer:

Element	Mass	% composition
Mg	9.03	9.03/12.51 x 100 = 72.2
N	3.48	3.48/12.51 x 100 = 27.8
Total	12.51	100%

The magnesium-based compound has the following composition: **Mg 72.2% and N 27.8%**

j. *A 27.0 g sample of a compound contains 7.20 g of C, 2.20 g of hydrogen and 17.6 g of oxygen. Calculate the percentage composition of the compound.*

Answer:

Mass of compound = 27 g

Element	Mass	% composition
C	7.2	7.2/ 27 x 100 = 26.7
H	2.2	2.2/ 27 x 100 = 8.1
O	17.6	17.6/ 27 x 100 = 65.2
Total		100%

The organic compound has the following composition: **C 26.7%, H 8.1% and O 65.2%**

Exercise 1.4

Determine the empirical formula from the following molecular formula:

a. C_6H_6

Answer:

Highest common denominator = 6
Empirical formula = **CH**

b. $C_2H_4O_2$

Answer:

Highest common denominator = 2
Empirical formula = CH_2O

c. P_4O_6

Answer:

Highest common denominator = 2
Empirical formula = P_2O_3

d. N_2O_5

Answer:

Highest common denominator = 1
Empirical formula = N_2O_5

e. C_6H_9

Answer:

Highest common denominator = 3
Empirical formula = C_2H_3

f. CH_2OHCH_2OH

Answer:

$CH_2OHCH_2OH = C_2H_6O_2$
Highest common denominator = 2
Empirical formula = CH_3O

g. $C_6H_8O_6$

   Answer:

   Highest common denominator = 2
   Empirical formula = $\mathbf{C_3H_4O_3}$

h. CuC_2O_4

   Answer:

   Highest common denominator = 1
   Empirical formula = $\mathbf{CuC_2O_4}$

i. Hg_2F_2

   Answer:

   Highest common denominator = 2
   Empirical formula = $\mathbf{HgF}$

j. N_2O_4

   Answer:

   Highest common denominator = 2
   Empirical formula = $\mathbf{NO_2}$

Exercise 1.5

a. *The empirical formula of a substance is CH_2O. What is its formula mass if it has a molecular mass of 180?*

   Answer:

   Relative formula mass of CH_2O = 12 + (1 x 2) + 16 = 30
   Molecular formula = $(CH_2O)_n$

$n = 180/30 = 6$

Molecular formula $= (CH_2O)_6 = \textbf{C}_6\textbf{H}_{12}\textbf{O}_6$

b. *Determine the molecular formula of chemical compound with a molecular mass of 46.0 and an empirical formula of NO_2.*

Answer:

Relative formula mass of $NO_2 = 14 + (16 \times 2) = 46$

Molecular formula $= (NO_2)_n$

$n = 46/46 = 1$

Molecular formula $= \textbf{NO}_2$

c. *Which chemical compound has the empirical formula PO_2 and a relative formula mass of 252?*

Answer:

Relative formula mass of $PO_2 = 31 + (16 \times 2) = 63$

Molecular formula $= (PO_2)_n$

$n = 253/63 = 4$

Molecular formula $= \textbf{P}_4\textbf{O}_8$

d. *What is the molecular formula of a compound with an empirical formula $C_2H_4O_4$ and a relative formula mass of 132?*

Answer:

Relative formula mass $C_2OH_4 = (2 \times 12) + 16 + (4 \times 1) = 44$

Molecular formula $= (C_2OH_4)_n$

$n = 132/44 = 3$

Molecular formula $= (C_2OH_4)_3 = \textbf{C}_6\textbf{O}_3\textbf{H}_{12}$

e. *A compound with an empirical formula of C_4H_4O and a relative formula mass of 272. What is the molecular formula of this compound?*

   Answer:

   Relative formula mass of C_4H_4O = (4 x 12) + (4 x 1) + 16 = 68
   Molecular formula = $(C_4H_4O)_n$
   n = 272/68 = 4
   Molecular formula = $(C_4H_4O)_4$ = $C_{16}H_{16}O_4$

f. *Determine the molecular formula of a compound with the empirical formula of CFBrO and a relative formula mass of 254.7.*

   Answer:

   Relative formula mass of CFBrO = 12 +19 +80 + 16 = 127
   Molecular formula = $(CFBrO)_n$
   n = 254/127 = 2
   Molecular formula = $(CFBrO)_2$ = $C_2F_2Br_2O_2$

g. *What is the molecular formula of a compound with an empirical formula of CH_2N and a molecular mass of 84?*

   Answer:

   Relative formula mass of CH_2N = 12 + (1 x 2) + 14 = 28
   Molecular formula = $(CH_2N)_n$
   n = 84/28 = 3
   Molecular formula = $(CH_2N)_3$ = $C_3H_6N_3$

h. *Dimethylglyoxime, has the empirical formula C_2H_4NO and a relative formula mass is 116. What is the molecular formula of the compound?*

Answer:

Relative formula mass of C_2H_4NO = (2 x 12) + (1 x 4) + 14 + 16 = 58
Molecular formula = $(C_2H_4NO)_n$
n = 116/58 = 2
Molecular formula = $(C_2H_4NO)_2$ = $C_4O_8N_2O_2$

i. *The common headache remedy, Ibuprofen has the empirical formula C_7H_9O and a molecular mass of approximately 218. Determine the molecular formula of Ibuprofen.*

Answer:

Relative formula mass of C_7H_9O = (7 x 12) + (1 x 9) + 16 = 109
Molecular formula = $(C_7H_9O)_n$
n = 218/109 = 2
Molecular formula = $(C_7H_9NO)_2$ = $C_{14}H_{18}O_2$

j. *Determine the molecular formula of a compound with an empirical formula of C_2H_3O and a molecular mass of 172.*

Answer:

Relative formula mass of C_2H_3O = (12 x 2) + (1 x 3) + 16 = 43
Molecular formula = $(C_2H_3O)_n$
n = 172/43 = 4
Molecular formula = $(C_2H_3NO)_4$ = $C_8H_{12}O_4$

Exercise 1.6

a. *A compound contains 24.74% potassium, 34.76% manganese, and 40.50% oxygen. Determine the empirical formula of this compound.*

Answer:

	Potassium	Manganese	Oxygen
% composition	24.7	34.76	40.5
A_r	19	55	16
% composition/A_r	24.7/19 = 1.3	34.76/55 = 0.63	53.4/16 = 3.4
ratio	2	1	7

Empirical formula: K_2MnO_4

b. *The percent composition of a compound was found to be 19.3% sodium, 26.9% sulfur, and 53.8% oxygen. Its formula mass is 238. What is its molecular formula?*

Answer:

	Sodium	Sulfur	Oxygen
% composition	19.3	26.9	53.8
A_r	19	55	16
% composition/A_r	19.3/19 = 1	26.9/55 = 0.49	53.4/16 = 3.4
ratio	2	1	7

Empirical formula: Na_2SO_7

c. *A compound contains 16.7 g of Iridium and 10.3 g of selenium, what is its empirical formula?*

Answer:

	Iridium	Selenium
% composition	16.7.27 = 61.85	10.3/27 = 38.15
A_r	192	119
% composition/A_r	61.85/192 = 0.322	38.15/119 = 0.32
ratio	1	1

Empirical formula: **IrSe**

d. *It is found that 207 g of lead combined with 32 g of sulphur to form 239 g of lead sulphide. From the data work out the formula of lead sulphide.*

Answer:

	Lead	Sulfur
% composition	207/239 = 86.6%	32/239 = 13.4
A_r	207	32
% composition/A_r	86.6/207 = 0.42	13.4/32 = 0.41
ratio	1	1

Empirical formula: **PbS**

e. *Find the empirical formula of a compound which has the following percent composition 41.3% C, 10.4% H, and 48.2% N.*

Answer:

	Carbon	Hydrogen	Nitrogen
% composition	41.3	10.4	48.2
A_r	12	1	14
% composition/A	41.3/12 = 3.44	10.4/1 = 10.4	48.2/14 = 3.44
ratio	1	3	1

Empirical formula: **CH_3N**

f. *Determine the empirical and molecular formula for a compound if its molar mass is 308 and contains 85.69% carbon; 3.93% hydrogen and 10.38% oxygen.*

Answer:

	Carbon	Hydrogen	Oxygen
% composition	85.69	3.93	10.38
A_r	12	1	16

Table contd.....

	Carbon	Hydrogen	Oxygen
% composition/A	85.69/12 = 7.14	3.93/1 = 3.93	10.38/16 = 0.64
ratio	11	6	1

Empirical formula: $C_{11}H_6O$

Relative formula mass $C_{11}H_6O = (11 \times 12) + (6 \times 1) + 16 = 154$

Molecular formula $= (C_{11}H_6O)_n$

$n = 308/154 = 2$

Molecular formula $= (C_{11}H_6O)_2 = \mathbf{C_{22}H_{12}O_2}$

g. *Determine the empirical and molecular formulas of the female steroid hormone, estradiol, which has a molecular mass of 278.*

Answer:

	Carbon	Hydrogen	Oxygen
% composition	77.63	10.88	11.49
A_r	12	1	16
% composition/A	77.63/12 = 6.47	10.88/1 = 10.88	11.49/16 = 0.72
ratio	9	15	1

Empirical formula: $C_9H_{15}O$

Relative formula mass $C_9H_{15}O = (9 \times 12) + (15 \times 1) + 16 = 139$

Molecular formula $= (C_9H_{15}O)_n$

$n = 278/139 = 2$

Molecular formula $= (C_9H_{15}O)_2 = \mathbf{C_{18}H_{30}O_2}$

h. *A sample of the amino acid glycine contains 1.56 g C, 0.333 g H, 2.08 g O, and 0.910 g N. Determine its empirical formula.*

Answer:

	Carbon	Hydrogen	Oxygen	Nitrogen
% composition	1.56/4.883 x 100 = 31.95	0.333/4.883 x 100 = 6.89	2.08/4.883 x 100 = 42.6	0.910/4.883 x 100 = 18.64
A_r	12	1	16	14

Table contd.....

	Carbon	Hydrogen	Oxygen	Nitrogen
% composition/A_r	31.95/12 = 2.66	6.89/1 = 6.89	42.6/16 = 2.66	118.64/14 = .33
ratio	2	5	2	1

Empirical formula: $C_2H_5O_2N_2$

i. *The composition of acetic acid is found to be 39.9% C, 6.7% H and 53.4% O. Determine the empirical formula of acetic acid.*

Answer:

	Carbon	Hydrogen	Oxygen
% composition	39.9	6.7	53.4
A_r	12	1	16
% composition/A_r	39.9/12 = 3.325	6.7/1= 6.7	53.4/16 = 3.33
Ratio	1	2	1

Empirical formula: CH_2O

j. *A 50.51g sample of a phosphorus chloride was decomposed. Analysis of the products showed that 11.39 g of phosphorus was produced. What is the empirical formula of the chloride?*

Answer:

	Phosphorus	Chlorine
% composition	11.39/50.51 x 100 = 22.55	39.12/50.51 x 100 = 77.45
A_r	31	35.5
% composition/A_r	22.35/31 = 0.72	77.45/35.5 = 2.2
Ratio	1	3

Empirical formula: PCl_3

k. *What is the empirical formula of a molecule containing 18.7% lithium, 16.3% carbon, and 65.0% oxygen?*

Answer:

	Lithium	Carbon	Oxygen
% composition	18.7	16.3	65
A_r	7	12	16
% composition/A_r	18.7/7 = 2.67	16.3/12 = 1.35	65/16 = 4.06
Ratio	2	1	3

Empirical formula: Li_2CO_3

l. *Use the following percent composition data to determine the empirical formula of an unknown compound: 40.50% Zn; 19.86% S; 39.64% O.*

Answer:

	Zinc	Sulfur	Oxygen
% composition	40.5	19.86	39.64
A_r	64	32	16
% composition/A_r	40.5/64 = 0.63	19.86/32 = 0.62	39.64/16 = 2.48
Ratio	1	1	4

Empirical Formula: $ZnSO_4$

m. *A hydrocarbon contains 40.10% carbon; 6.49% hydrogen and 51.41% oxygen. Determine the empirical formula of the compound.*

Answer:

	Carbon	Hydrogen	Oxygen
% composition	40.1	6.49	51.41
A_r	12	1	16
% composition/A_r	40.1/12 = 3.34	6.49/1 = 6.49	51.41/16 = 3.21
Ratio	1	2	1

Empirical formula: CH_2O

n. *A nitrogen oxide contains 36.84% N. What is the empirical formula of the oxide?*

Answer:

	Nitrogen	Oxygen
% composition	36.84	63.16
A_r	14	16
% composition/A_r	36.84/14 = 2.63	63.16/16 = 3.94
Ratio	1 2	1.5 3

Empirical formula: N_2O_3

o. *A salt was found to have the following percent composition: 47.0 % potassium, 14.5 % carbon, and 38.5 % oxygen. What is its empirical formula? If the compound has a molecular mass of 166.22, what is its molecular formula?*

Answer:

	Potassium	Carbon	Oxygen
% composition	47.0	14.5	38.5
A_r	39	12	16
% composition/A_r	47/39 = 1.2	14.5/12 = 1.2	38.5/16 = 2.4
Ratio	1	1	2

Empirical formula: KCO_2
Relative formula mass KCO_2 = (1 x 39) + 12 + (2 x 16) = 83
Molecular formula = $(KCO_2)_n$
n = 166/83 = 2
Molecular formula = $(KCO_2)_2$ = $K_2C_2O_4$

CHAPTER 2

The Mole

Keywords: Avogadro's number, Calculations, Formula units, Molar mass, Relative formula mass, The mole.

The Mole is simply a number. Just as the term dozen refers to the number (12) twelve and a score to the number (20) twenty, the mole refers to the number 6.023 x 10^{23}. Thus 12 eggs is a dozen of eggs, 20 eggs is a score of eggs and 6.023 x 10^{23} eggs is a mole of eggs. Commonly referred to as Avogadro's constant, 6.023 x 10^{23} is the number of atoms found in exactly 12 grams of carbon-12. Carbon-12 is used as the standard from which atomic masses are measured: its mass number is 12 by definition. Since 12 g of carbon contains one mole of carbon atoms, the mass of one mole of any element is equal to its relative atomic mass in grams. Magnesium has relative atomic mass of 24. Therefore one mole of magnesium has a mass of 24 g. Thus 24 g of magnesium contains 6.02 x 10^{23} magnesium atoms and the mass of one atom of magnesium = 24/(6.02 x 10^{23}) = 3.987 x 10^{-23} g. Similarly, the mass of one mole of lithium (A_r = 7) is 7g, 27 g of aluminium (A_r = 27) contains one mole of atoms and the mass of one mole of calcium (A_r = 40) is 40g *etc.* You can also work with fractions (or multiples) of moles:

Mole/Mass Relationship Examples Using Magnesium (A_r =24)		
Moles Magnesium	**Number of Magnesium Atoms**	**Mass of Magnesium**
0.25	1.505 x 10^{23}	6 g
0.5	3.01 x 10^{23}	12 g
1	6.02 x 10^{23}	24 g
2	1.204 x 10^{24}	48 g
10	6.02 x 10^{24}	240 g
50	3.01 x 10^{25}	1200 g

Some elements exist as molecules rather than atoms. The following elements all exist as diatomic molecules: hydrogen (H_2), nitrogen (N_2), oxygen (O_2) and the

Nigel P. Freestone
All rights reserved-© 2016 Bentham Science Publishers

halogens (F_2, Cl_2, Br_2, I_2). Hydrogen has a relative atomic mass of 1. Therefore, the relative formula mass of (M_r) of H_2 = (2 x 1) = 2. Therefore, one mole of hydrogen molecules will have a mass of 2g and will cont 6.02×10^{23} molecules of hydrogen. Oxygen has a relative atomic mass of 16. Thus one mole of oxygen gas (O_2) has a mass of 32 g and 6.02×10^{23} molecules of nitrogen gas (N_2) have a mass of 28g.

The concept of a mole is equally applicable to compounds as well as elements. The mass of one mole of a compound is its relative formula mass (M_r) in grams. To avoid any ambiguity it is convenient to use the term **formula unit**. Formula unit refers to the smallest repeating unit of a substance and is the chemical formula normally used for the substance. For instance, the formula unit of graphite is an atom of carbon (C). Similarly, the formula unit of oxygen gas is an oxygen molecule (O_2); NaCl is the formula unit for the ionic compound sodium chloride and the formula unit for silicon dioxide is SiO_2.

Equimolar amounts of substances contain the same number of formula units. Thus 0.5 moles any substance will contain the same number of formula units (particles), *i.e.* $0.5 \times 6.02 \times 10^{23} = 3.01 \times 10^{23}$.

The idea of the **mole** links the **mass of a substance** to the **number of formula units** (particles) it contains. The mass of one mole of an element or compound is referred to as its **molar mass**, which is its **relative atomic mass (A_r)** or **relative formula mass(M_r)** in grams.

Molar Mass (M_r) = Relative Formula Mass in grams (g mol^{-1})

If you have *m* grams of a substance which has a molar mass of M_r g mol^{-1}, then the amount of a substance in moles, *n*, is given by:-

Number of moles = Mass (g) / Molar Mass (g mol^{-1})
Number of formula units (particles) = Number of moles x 6.23×10^{23}
Number of moles = mass (*g*) / M_r (g mol^{-1})

Thus if you know the values of any two of n, m or M_r you can calculate the third using the equations above.

Water has a relative formula mass of 18. Thus:

- one mole of water has a mass of 18 g;
- 18 g of water contains 6.02 x 10^{23} formula units (molecules) of water;
- 0.5 moles of water has a mass of 9 g and contains 3.01 x 10^{23} molecules of water;
- one molecule of water has a mass of $18/(6.02 \times 10^{23}) = 2.99 \times 10^{-23}$ g.

Example 2.1: *Determine the mass of one mole of O_2?*

Answer:

Relative formula mass of O_2 = (2 x 16) = 32
Mass of one mole, *i.e.* molar mass = M_r in g
Molar Mass, $M_r O_2$ = **32 g mol^{-1}**

Example 2.2: *What is the mass of 0.05 moles of ammonium sulfate?*

Answer:

Relative formula mass of $(NH_4)_2SO_4$ = (2 x 14) + (8 x 1) + 32 + (4 x 16) = 132
Mass of one mole, *i.e.* molar mass = M_r in g
Molar Mass, M_r $(NH_4)_2SO_4$ = 132 g mol^{-1}
Therefore, 0.05 moles of ammonium sulfate has a mass of 132 x 0.05 = **6.6 g**

Example 2.3: *How many moles of substance are present in 0.250 g of calcium carbonate?*

Answer:

Relative formula mass of $CaCO_3$ = 40 + 12 + (3 x 16) =100
Mass of one mole, *i.e.* molar mass = M_r in g
Molar Mass, M_r $CaCO_3$ = 100g mol^{-1}
So the number of moles in 0.250 g of $CaCO_3$ = mass/M_r = 0.250/100 = **2.5 x 10^{-3} moles**

Example 2.4: *How many formula units are present in 9 g of KNO_3?*

Answer:

Relative formula mass of KNO_3 = 39 + 14 + (3 x16) = 101
Mass of one mole, *i.e.* molar mass = M_r in g
Molar Mass, Mr (KNO_3) = 101 g
Number of moles in 9 g of KNO_3 = mass/M_r = 9/101 = 0.09
One mole of a substance contains 6.02 x 10^{23} formula unit particles
Therefore, the number of formula particle units in 0.09 moles = number of moles x 6.02 x 10^{23} = 0.09 x 6.02 x 10^{23} = **45.41 x 10^{22}**

Example 2.5: *An average person's respiration generates approximately 37.5 g of carbon dioxide per hour. How many molecules are in 37.5 g of carbon dioxide (CO_2)?*

Answer:

M_r CO_2 = 44 g mol^{-1}
Number of moles of CO_2 in 37.5 g = mass/M_r = 37.5/44 = 0.85 moles
One mole of CO_2 contains 6.02 x 10^{23} molecules of CO_2
Therefore, number of molecules in 0.85 moles of CO_2 = number of moles x 6.02 x 10^{23} = 0.85 x 6.02 x 10^{23} = **5.12 x 10^{23}**

Example 2.6: *What mass of ozone (O_3) contains 3.67 x 10^{22} molecules of O_3?*

Answer:

M_r O_3 = 48 g mol^{-1}
Thus 48 g of O_3 contains 6.02 x 10^{23} molecules of ozone
One molecule of O_3 will have a mass of 48 / (6.02 x 10^{23})
Therefore, 3.67 x 10^{22} molecules of O_3 will have a mass of 48 / (6.02 x 10^{23}) x 3.67 x 10^{22} =**2.93 g**

Example 2.7: *Complete the following table relating to calcium carbonate*

Substance	M_r	Number of moles (n)	Mass in grams (m)	Number of particles
Carbon dioxide, CO_2	i)	1.5	vii)	x)

Table contd.....

Substance	M_r	Number of moles (n)	Mass in grams (m)	Number of particles
Nitrogen, N_2	ii)	v)	7	xi)
Sulfur Dioxide, SO_2	iii)	0.15	viii)	xii)
Ethanol, C_2H_5OH	iv)	vi)	ix)	1.2×10^{21}

Answer:

Substance	M_r	Number of moles (n)	Mass in grams (m)	Number of particles
Carbon dioxide, CO_2	44	1.5	66	9.03×10^{23}
Nitrogen, N_2	28	0.25	7	1.505×10^{23}
Sulfur Dioxide, SO_2	64	0.15	9.6	9.03×10^{22}
Ethanol, C_2H_5OH	46	0.002	0.09	1.2×10^{21}

M_r values

   i. $M_r\ CO_2 = 12 + (2 \times 16) = \mathbf{44}$
  ii. $M_r\ N_2 = (2 \times 14) = \mathbf{28}$
 iii. $M_r\ SO_2 = 32 + (2 \times 16) = \mathbf{64}$
 iv. $M_r\ C_2H_5OH = (2 \times 12) + (5 \times 1) + 16 + 1 = \mathbf{46}$

Number of moles
Number of formula units (particles) = Number of moles x 6.23 x 10^{23}

  v. Number of moles in 7g of N_2 = mass/M_r = 7/28 = **0.25**
 vi. Number of moles in 1.2 x 10^{21} particles of C_2H_5OH = number of particles/6.02
    x 10^{23} = 1.2 x 1021/6.02 x 10^{23} = **0.002**

Mass in grams
Mass (g) = Number of moles x M_r

 vii. Mass of 1.5 moles of CO_2 = number of moles x M_r = 1.5 x 44 = **66 g**
viii. Mass of 0.15 moles SO_2 = number of moles x M_r =0.15 x 64 = **9.6 g**

Number of particles
Number of formula units (particles) = Number of moles x 6.23 x 10^{23}

ix. Number of particles in 1.5 moles CO_2 = number of moles x 6.02 x 10^{23} = **9.03 x 10^{23}**

x. Number of moles in 7g of N_2 = mass/M_r = 7/28 = **0.25**

xi. Number of particles in 0.25 moles of CO_2 = number of moles x 6.02 x 10^{23} = 0.25 x 6.02 x 10^{23} =**1.505 x 10^{23}**

xii. Number of particles in 0.15 moles of SO_2 = number of moles x 6.02 x 10^{23} = **9.03 x 10^{22}**

In summary, a mole always contains the same number of formula units (particles) regardless of the substance. But, the mass of a mole differs from substance to substance, and is the relative formula mass expressed in grams. Really the mole is just a collective term like the dozen. A dozen elephants weigh more than a dozen mice, but we have the same number of each.

It is very important to state the particles you are referring to when talking about moles. A mole of oxygen could refer to a mole of oxygen atoms or to a mole of oxygen gas, which is diatomic (O_2). So a mole of oxygen atoms (O) will have a mass of 16g, while a mole of oxygen gas (O_2) has a mass of 16 x 2 = 32g.

The concept of the mole is useful because the size and mass of atoms are so small; hence Avogadro's number is so large.

If this makes sense then you have already started **Counting Moles ...**

Exercise 2.1

Calculate the molar masses (M_r) of the following:-

a. *Chlorine, Cl_2*
b. *Sulfur dioxide, SO_2*
c. *Zinc nitrate, $Zn(NO_3)_2$*
d. *Magnesium carbonate, $MgCO_3$*
e. *Oxalic acid, $C_2H_4O_2$*
f. *Calcium chloride, $CaCl_2$*
g. *Aluminium sulphate, $Al_2(SO_4)_3$*

h. *Sulfuric acid, H_2SO_4*
i. *Potassium mangante (VII), $KMnO_4$*
j. *Sodium chromate (VI), Na_2CrO_4*

Exercise 2.2

How many moles of substance are present in the following?

a. *5.30 g of sodium carbonate, Na_2CO_3*
b. *0.35 g of zinc nitrate, $Zn(NO_3)_2$*
c. *0.008 g of sodium hydroxide, $NaOH$*
d. *1.25 g of calcium carbonate, $CaCO_3$*
e. *3.5 g of benzene, C_6H_6*
f. *12 g of glucose, $C_6H_{12}O_6$*
g. *1g of uranium dioxide, UO_2*
h. *0.3 g aluminium sulphate, $Al_2(SO_4)_3$*
i. *1.2 g iron (III) oxide, Fe_2O_3*
j. *3.4 g sulphur trioxide, SO_3*

Exercise 2.3

How many formula units are present in the following?

a. *0.25 moles of Cl_2*
b. *5 moles of CO_2*
c. *10g of $CaCO_3$*
d. *2.45 x 10^3 moles of NH_3*
e. *0.34 kg of Fe_3O_4*
f. *2.56 moles of C_6H_6*
g. *1 x 10^6 g of Au*
h. *0.12 moles of $CuSO_4$*
i. *1 tonne of N_2*
j. *4.45 x 10^6 moles of $(NH_4)_2CO_3$*

Exercise 2.4

Determine the mass of the following:

a. *2 moles of carbon dioxide, CO_2*
b. *0.01 moles of nitrogen dioxide, NO_2*
c. *1×10^5 moles of benzene, C_6H_6*
d. *2.03×10^3 moles of uranium dioxide, UO_2*
e. *1.12 moles of sulphuric acid, H_2SO_4*
f. *3×10^4 moles of calcium carbonate, $CaCO_3$*
g. *1.2 moles of ethane, C_2H_4*
h. *0.5 moles ethanoic acid, CH_3COOH*
i. *1.25×10^3 moles sodium hydroxide, $NaOH$*
j. *0.025 moles potassium dichromate, $K_2Cr_2O_7$*

ANSWERS

Exercise 2.1

Calculate the molar masses (M_r) of the following:-

Molar Mass = Relative Formula Mass (M_r) in g

Answers

a. $Cl_2 = (35.5 \times 2) = $ **71 g**
b. $SO_2 = 32 + (16 \times 2) = $ **64 g**
c. $Zn(NO_3)_2 = 65 + \{2 \times (14 + (16 \times 3))\} = $ **189 g**
d. $MgCO_3 = 24 + 12 + (16 \times 3) = $ **84 g**
e. $C_2H_4O_2 = (2 \times 2) + (1 \times 4) + (16 \times 2) = $ **60 g**
f. $CaCl_2 = 40 + (35.5 \times 2) = $ **111g**
g. $Al_2(SO_4)_3 = (27 \times 2) + \{3 \times (32 + (16 \times 4))\} = $ **342 g**
h. $H_2SO_4 = \{(1 \times 2) + 32 + (16 \times 4)\} = $ **98 g**
i. $KMnO_4 = \{39 + 55 + (16 \times 4)\} = $ **158 g**
j. $Na_2CrO_4 = \{(23 \times 2) + 52 + (16 \times 4)\} = $ **162 g**

Exercise 2.2

How many moles of substance are present in the following?

Number of Moles = Mass/M_r

a. *5.30 g of sodium carbonate, Na_2CO_3*

   Answer:

   M_r $Na(CO_3)_2$ = 106 g mol^{-1}
   Number of moles in 5.3 g of sodium carbonate = mass/M_r = 5.3/106 = **0.05**

b. *0.35 g of zinc nitrate, $Zn(NO_3)_2$*

   Answer:

   M_r $Zn(NO_3)_2$ = 189 g mol^{-1}
   Number of moles in 0.35 g of zinc nitrate = mass/M_r = 0.35/189 = **1.85 x 10^{-3}**

c. *0.008 g of sodium hydroxide, NaOH*

   Answer:

   M_r NaOH = 40 g mol^{-1}
   Number of moles in 0.008g of sodium hydroxide = mass/M_r = 0.008/40 = **2 x 10^{-4}**

d. *1.25 g of calcium carbonate, $CaCO_3$*

   Answer:

   M_r $CaCO_3$ = 100 g mol^{-1}
   Number of moles in 1.25g of calcium carbonate = mass/M_r = 1.25/100 = **0.0125**

e. *3.5 g of benzene, C_6H_6*

   Answer:

   M_r C_6H_6 = 78 g mol^{-1}
   Number of moles of benzene in 3.5g = mass/M_r = 3.5/78 = **0.045**

f. *12 g of glucose, $C_6H_{12}O_6$*

   Answer:

   M_r $C_6H_{12}O_6$ = 180 g mol^{-1}
   Number of moles in 12g of glucose = mass/M_r = 12/180 = **0.067**

g. *1g of uranium dioxide, UO_2*

   Answer:

   M_r UO_2 = 270 g mol^{-1}
   Number of moles in 1g of uranium dioxide = mass/M_r = 1/270 = **3.7 x 10^{-3}**

h. *0.3 g aluminium sulphate, $Al_2(SO_4)_3$*

   Answer:

   M_r $Al_2(SO_4)_3$ = 342 g mol^{-1}
   Number of moles in 0.3g aluminium sulphate = mass/M_r = 0.3/342 = **8.77 x 10^{-4}**

i. *1.2 g iron (III) oxide, Fe_2O_3*

   Answer:

   M_r Fe_2O_3 = 160 g mol^{-1}
   Number of moles in 1.2 g iron (III) oxide = mass/M_r = 1.2/160 = **7.5 x 10^{-3}**

j. *3.4 g sulphur trioxide, SO_3*

Answer:

M_r SO_3 = 80g mol^{-1}

Number of moles in 3.4 g sulphur trioxide, SO_3 = mass/M_r = 3.4 /80 = **0.0425**

Exercise 2.3

How many formula units are present in the following?

Number of formula units = Number of moles x 6.02 x 10^{23}

a. *0.25 moles of Cl_2*

Answer:

Number of molecules of Cl_2 in 0.25 moles = 0.25 x 6.02 x 10^{23} = **1.51 x 10^{23}**

b. *5 moles of CO_2*

Answer:

Number of molecules of CO_2 in 5 moles = 5 x 6.02 x 10^{23} = **3.01 x 10^{24}**

c. *10g of $CaCO_3$*

Answer:

M_r $CaCO_3$ = 100

Number of moles in 10g of $CaCO_3$ = mass/Mr = 10/100 = 0.1

Number of formula units in 10 g of $CaCO_3$ = 0.1 x 6.02 x 10^{23} = **6.02 x 10^{22}**

d. *2.45 x 10^3 moles of NH_3*

Answer:

Number of formula units (molecules) of NH_3 in 2.45 x 10^{-3} moles = 2.45 x 10^{-3}

x 6.02 x 10^{23} = **1.48 x 10^{21}**

e. *0.34 kg of Fe_3O_4*

   Answer:

   M_r Fe_3O_4 = 232
   Number of moles in 340 g (0.34 kg) of Fe_3O_4 = mass/M_r = 340/232 = 1.47
   Number of formula units in 0.24 kg of Fe_3O_4 = 1.47 x 6.02 x 10^{23} = **8.85 x 10^{23}**

f. *2.56 moles of C_6H_6*

   Answer:

   Number of particles (molecules) in 2.56 moles of C_6H_6 = 2.56 x 6.02 x 10^{23} =
   1.54 x 10^{24}

g. *1 x 10^{-6} g of Au*

   Answer:

   Number of moles in 1 x 10^{-6} g of Au = 1 x 10^{-6}/197 = 5.08 x 10^{-9}
   Number of atoms in 1 x 10^{-6} g of Au = 5.08 x 10^{-9} x 6.02 x 10^{23} = **3.06 x 10^{15}**

h. *0.12 moles of $CuSO_4$*

   Answer:

   Number of formula units of $CuSO_4$ in 0.12 moles = 0.12 x 6.02 x 10^{23} = **7.22 x
   10^{22}**

i. *1 tonne of N_2*

   Answer:

   M_r N_2 = 28

Number of moles of 1000 g (1 tonne) in N_2 = 1000/28 = 35.7
Number of formula units in 1 tonne of N_2 = 35.7 x 6.02 x 10^{23} = **2.15 x 10^{25}**

j. *4.45 x 10^{-6} moles of $(NH_4)_2CO_3$*

Answer:

Number of formula units in 4.45 x 10^{-6} moles of $(NH_4)_2CO_3$ =4.45 x 10^{-6} x 6.02 x 10^{23} = **2.68 x 10^{18}**

Exercise 2.4

Determine the mass of the following:

Mass = Number of Moles x M_r

a. *2 moles of carbon dioxide, CO_2*

Answer:

M_r CO_2 = 44 g mol^{-1}
Mass of 2 moles of carbon dioxide = number of moles x M_r = 2 x 44 = **88 g**

b. *0.01 moles of nitrogen dioxide, NO_2*

Answer:

M_r NO_2 = 48 g mol^{-1}
Mass of 0.01 moles of nitrogen dioxide = number of moles x M_r = 0.01 x 46 = **0.46 g**

c. *1 x 10^{-5} moles of benzene, C_6H_6*

Answer:

M_r C_6H_6 = 78 g mol^{-1}
Mass of 1 x 10^{-5} moles of benzene = number of moles x M_r = 1 x 10^{-5} x 78 = **7.8 x 10^{-4} g**

d. *2.03 x 10^{-3} moles of uranium dioxide, UO_2*

Answer:

M_r UO_2 = 270 g mol^{-1}
Mass of 2.03 x 10^{-3} moles of uranium dioxide = number of moles x M_r = 2.03 x 10^{-3} x 270 = **0.55 g**

e. *1.12 moles of sulphuric acid, H_2SO_4*

Answer:

M_r H_2SO_4 = 98 g mol^{-1}
Mass of 1.12 moles of sulfuric acid = number of moles x M_r = 1.12 x 98 = **109.76 g**

f. *3 x 10^{-4} moles of calcium carbonate, $CaCO_3$*

Answer:

M_r $CaCO_3$ = 100 g mol^{-1}
Mass of 3 x 10^{-4} moles of calcium carbonate= number of moles x M_r = 3 x 10^{-4} x 100 = **0.03 g**

g. *1.2 moles of ethane, C_2H_4*

Answer:

M_r C_2H_4 = 28 g mol^{-1}
Mass of 1.2 moles of ethane = number of moles x M_r = 1.2 x 28 = **33.6 g**

h. *0.5 moles ethanoic acid, CH_3COOH*

Answer:

M_r CH_3COOH = 60 g mol^{-1}

Mass of 0.5 moles ethanoic acid = number of moles x M_r = 0.5 x 60 = **30 g**

i. *1.25 x 10^3 moles sodium hydroxide, NaOH*

Answer:

M_r NaOH = 40 g mol^{-1}
Mass of 1.25 x 10^{-3} moles of sodium hydroxide = number of moles x M_r =
1.25 x 10^{-3} x40 = **0.05 g**

j. *0.025 moles potassium dichromate, $K_2Cr_2O_7$*

Answer:

M_r $K_2Cr_2O_7$ = 294 g mol^{-1}
Mass of 0.025 moles of potassium dichromate = number of moles x M_r =
0.025 x 294 = **7.35 g**

CHAPTER 3

Moles & Balanced Chemical Equations

Keywords: Balancing chemical equations moles, Calculations, Calculating frame, Chemical equations, Products, Reactants, Reaction coefficients, Stoichiometry.

3.1. CHEMICAL EQUATIONS

Chemical equations are a form of shorthand that scientists use to describe chemical reactions, where the reactants are given on the left-hand side and the product on the right hand side.

$$Reactants \rightarrow Products$$

Chemical reactions simply involve the rearrangement of the same atoms through the breaking of chemical bonds and the formation of new bonds. Equations are balanced when the number and type of reactant and product atoms are identical. To balance equations, whole numbers, known as **reaction coefficients** are placed to the left hand side of chemical formula. **Remember** you cannot change the chemical formula of a substance to balance a chemical equation. Mass is always conserved. Atoms of gold cannot be produced from atoms of lead.

Sodium carbonate breaks down when heated to produce sodium carbonate, carbon dioxide and water. This thermal degradation can be represented by the following word equation:

sodium hydrogen carbonate $\rightarrow$ sodium carbonate + carbon dioxide + water

Although this word equation is a convenient way of expressing a chemical reaction, it really does not provide us with any more information than the opening sentence of the paragraph above. But this all changes when we replace the names of the chemicals with chemical formula.

$$NaHCO_3 \rightarrow Na_2CO_3 + CO_2 + H_2O$$

Nigel P. Freestone
All rights reserved-© 2016 Bentham Science Publishers

Atom Count (unbalanced):

	Reactants	Products
Na atoms	1	2
H atoms	1	2
C atoms	1	2
O atoms	3	6

The above reaction is unbalanced since the number of atoms of each element on the left hand side of the equation is not the same as the number of atoms of each element on the right hand side. To balance the equation the relative number of molecules or formula units that participate in the reaction are changed. Reaction coefficients are placed to the left of the chemical formula, *e.g.*, $2NaHCO_3$, until the number of atoms of each element on either side of the equation balance. For most chemical equations it is often best to start by balancing the element with the least atoms present.

Once one element is balanced, proceed to balance another, and another, until all elements are balanced. Thus for the reaction above, we can see that H appears in one reactant ($NaHCO_3$) and one product (H_2O). H's can be balanced by using '2' as the coefficient for $NaHCO_3$.

$$2NaHCO_3 \rightarrow Na_2CO_3 + CO_2 + H_2O$$

Atom Count (balanced):

	Reactants	Products
Na atoms	2	2
H atoms	2	2
C atoms	2	2
O atoms	6	6

As you can see from the atom count, the equation is now balanced. Equations are completed by indicating the states of matter of each chemical species present, *i.e.* - (g) for gaseous substances, (s) for solids, (l) for liquids, and (aq) for species in solution in water.

$$2NaHCO_{3\,(s)} \rightarrow Na_2CO_{3(s)} + CO_{2(g)} + H_2O_{\,(l)}$$

Example 3.1: *Balance the following equations:*

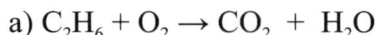

a) $C_2H_6 + O_2 \rightarrow CO_2 + H_2O$

Start by balancing the element with the least atoms present, *i.e.* C, then the H atoms and lastly the O atoms.

     Step 1: Balance C's

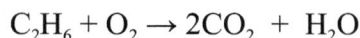

$$C_2H_6 + O_2 \rightarrow 2CO_2 + H_2O$$

     Step 2: Balance H's

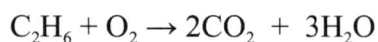

$$C_2H_6 + O_2 \rightarrow 2CO_2 + 3H_2O$$

     Step 3: Balance O's

$$C_2H_6 + 7/2\ O_2 \rightarrow 2CO_2 + 3H_2O$$

Although the equation is balanced in terms of the atom count, the reaction coefficients need to be the lowest possible whole numbers. To get whole numbers we multiply all reaction coefficients by the lowest common denominator of any fractions. In this example we multiply by 2 to remove the 7/2 fraction.

    Balanced equation: **$2C_2H_6 + 7O_2 \rightarrow 2CO_2 + 3H_2O$**

b) $N_2 + H_2 \rightarrow NH_3$

Start by balancing the element with the least atoms present, *i.e.* N.

    Step 1: Balance the N's. Since there are 2 N's on the left hand side of the equation and one on the right had side, the N's can be balanced by using the coefficient 2 in front of the NH_3.

$$N_2 + H_2 \rightarrow 2NH_3$$

    Step 2: Balance the H's. Since there are 3 H's on the left hand side and 6 on the right hand side, the H's can be balanced by placing the coefficient 3 in front of H_2.

$$N_2 + 3H_2 \rightarrow 2NH_3$$

Balanced equation: $\mathbf{N_2 + 3H_2 \rightarrow 2NH_3}$

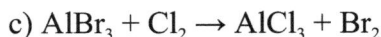

c) $AlBr_3 + Cl_2 \rightarrow AlCl_3 + Br_2$

Start by balancing the element with the least atoms present, *i.e.* Al, followed by Cl and lastly Br.

Step 1:   Balance Al. There is one Al on both sides of the equation – so Al is already balanced.

Step 2:   Balance Cl. There are 2 Cl atoms on the left hand side of the equation and three on the right hand side. The lowest common denominator of the 2 and 3 (the subscripts of Cl containing formula) is 6. Therefore for there to be 6Cl's on either side of the equation.

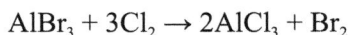

$$AlBr_3 + 3Cl_2 \rightarrow 2AlCl_3 + Br_2$$

The Al needs to be balanced again, *i.e.*

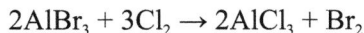

$$2AlBr_3 + 3Cl_2 \rightarrow 2AlCl_3 + Br_2$$

Step 3:   Balance Br. There are 6Br's on the left hand side and 3 Br's on the right hand side. So this can be easily placed by using the coefficient '3' in front of Br_2.

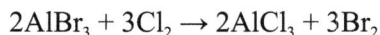

$$2AlBr_3 + 3Cl_2 \rightarrow 2AlCl_3 + 3Br_2$$

Balanced equation: $\mathbf{2AlBr_3 + 3Cl_2 \rightarrow 2AlCl_3 + 3Br_2}$

As with most things in life, the more your practice, the easier it becomes!

Exercise 3.1

Balance the following chemical equations

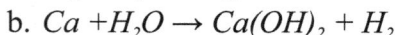

a. $Mg + O_2 \rightarrow MgO$
b. $Ca + H_2O \rightarrow Ca(OH)_2 + H_2$

c. $CuCO_3 + H_2SO_4 \rightarrow CuSO_4 + H_2O + CO_2$

d. $CH_4 + H_2O \rightarrow CO_2 + H_2$

e. $NO_2 + H_2O \rightarrow HNO_3 + NO$

f. $NaCl + H_2O \rightarrow NaOH + Cl_2 + H_2$

g. $CaCl_2 + Na_2CO_3 \rightarrow CaCO_3 + NaCl$

h. $C_{12}H_{22}O_{11} + O_2 \rightarrow CO_2 + H_2O$

i. $(NH_4)_2CO_3 \rightarrow NH_3 + CO_2 + H_2O$

j. $Fe + Cl_2 \rightarrow FeCl_3$

k. $Fe + O_2 \rightarrow Fe_2O_3$

l. $Ba(OH)_2 + H_3PO_4 \rightarrow H_2O + Ba_3(PO_4)_2$

m. $N_2O_5 \rightarrow NO_2 + O_2$

n. $CaCl_2 + AgNO_3 \rightarrow AgCl + Ca(NO_3)_2$

o. $Na_2SO_4 + C + CaCO_3 \rightarrow Na_2CO_3 + CaS + CO_2$

And finally:

$$CuSCN + KIO_3 + HCl \rightarrow CuSO_4 + KCl + HCN + ICl + H_2O$$

If you can balance this equation then you have mastered the black art of equation balancing!

3.2. MOLES CALCULATING FRAME

Moles have many uses in chemistry, including determining the amounts of substances used and produced in chemical reactions, and expressing strengths of solutions such as acids. The Counting Moles approach to solving mole-based problems utilises the '**Moles Calculating Frame**', which is both intuitive and simple to use.

Introducing the 'Moles Calculating Frame'

The 'Moles Calculating Frame' provides a simple and logical approach to solving all chemistry calculations involving equations:

	A	**+**	**2B**	**→**	**2C**	**+**	**3D**
M_r							
Mass Balance							

Table contd.....

	A	**+**	**2B**	**→**	**2C**	**+**	**3D**
Reaction Coefficients	1		2		2	:	3
Mass							
Number of Moles							
Concentration (M)							
Volume							

simply:

- construct the frame around the balanced chemical equation;
- in the left hand column list the parameters – *i.e.* molar mass, mass, number of moles, reaction coefficients *etc*;
- insert all given information. Remember you will always be able to determine M_r, and reaction coefficients;
- use a question mark to identify which parameter you need to calculate to answer the question;
- identify which gaps in the framework you can calculate from the given information.

Useful equations

$$\text{Mass = Number of moles x } M_r$$
$$\text{Number of moles = Mass / } M_r$$
$$\text{Number of moles = Volume /Concentration}$$
$$\text{Concentration = Number of moles / Volume}$$
$$\text{Volume = mass (g) x } V_m$$

Units: mass in grams, volume in litres (L), M_r in g mol^{-1} and concentration in M.

Note: remember to take the reaction coefficients *i.e.* stoichiometry into account.

The Moles Calculating Frame is highly flexible and by changing the parameters it can be used to solve all moles-based chemistry problems.

Now let's apply the '**Mole Calculating Frame**' to solve Mole-based problems.

3.3. MOLE, MASSES AND CHEMICAL EQUATIONS

Chemical equations convey a vast amount of information. Consider the combustion of hydrogen gas in air to form water:

$$2H_{2(g)} + O_{2(g)} \rightarrow 2H_2O_{(l)}$$

	$2H_{2(g)}$	+	$O_{2(g)}$	$\rightarrow$	$2H_2O_{(l)}$
M_r	2		32		18
Reaction Coefficients	2		1		2
Mass(g)	2x2=4		32		2x18=36

The balanced equation above tells us that:

- 2 molecules of hydrogen gas react one with molecule of oxygen to produce two molecules of liquid water;
- 2 moles of hydrogen gas react with one mole of oxygen gas to produce 2 moles of liquid water;
- 4 g of H_2 react with 32 g of O_2 to produce 36g of H_2O;
- 1 g of H_2 reacts with 8 g of O_2 to produce 9 g of H_2O;
- 1 g of H_2O is produced from the oxidation of 4/36 = 0.11 g of H_2.

As we shall see the reaction coefficients in balanced chemical equations can be used to determine the relative number of moles (molecules, formula units *etc*) of a compound involved in a chemical reaction.

The combustion of natural gas (methane) can be represented by the following equation:

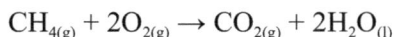

$$CH_{4(g)} + 2O_{2(g)} \rightarrow CO_{2(g)} + 2H_2O_{(l)}$$

This balanced equations tells us:

- 1 molecule of methane reacts with 2 molecules of oxygen to produce 1 molecule of carbon dioxide and two molecules of water;

- since one mole equates to 6.023×10^{23} particles, we can also state that, 1 mole of methane reacts with 2 moles of oxygen to produce 1 mole of carbon dioxide and two moles of water;
- the relative quantities of reactants and products indicated by the reaction coefficients is often referred as the **reaction stiochiometry.**

Given the relative atomic masses (H = 1, C = 12, O = 16), we can determine the M_r for each species in the equation and ensure that the equation is balanced by performing a mass balance.

	$CH_{4(g)}$	+	$2O_{2(g)}$	$\rightarrow$	$CO_{2(g)}$	+	$2H_2O_{(l)}$
M_r	16		32		44		18
Mass Balance	16		2x32=64		44		2x18=36
		80				80	

Note: the sum of the masses of the reactants = the masses of the products. A balanced chemical equation is thus simply a mass (and atom) balance exercise!

Thus 16 g of CH_4 requires 64 g of O_2 for complete combustion, forming 44 g of CH_4 and 36 g (or kg, tonnes *etc*) of H_2O. Thus 1 kg of CH_4 will generate 2.75 kg (44/16) CO_2 ...*etc.*

Note, the unit of quantity is irrelevant, we could discuss kilograms or tonnes or pounds for that matter, so long as we are consistent.

Example 3.2: *Use the following chemical equation to answer the questions below:*

$$2H_{2(g)} + 1O_{2(g)} \rightarrow 2H_2O_{(l)}$$

a) How many moles of oxygen will need to be consumed to produce 8 moles of water ?

Answer:

Construct a mole calculating frame around the balanced equation.

	$2H_{2(g)}$	+	$O_{2(g)}$	→	$CO_{2(g)}$
A_r/M_r	2		32		18
Mass Balance	4		32		2x18=36
		36			
Reaction Coefficients	2		1		2
Number of Moles	8		8/2=4		8

Thus 1 mole H_2O is produced from 0.5 mole of O_2

Therefore, 8 moles of H_2O are produced from 0.5 x 8 = **4 moles of O_2**

b) What mass of water will be produced from the combustion of 100g of hydrogen ?

Answer:

	$2H_{2(g)}$	+	$O_{2(g)}$	→	$CO_{2(g)}$
A_r/M_r	2		32		18
Mass Balance	4		32		2x18=36
		36			
Reaction Coefficients	2		1		2
Mass(g)	100				50x 18=**900**
Number of Moles	100/2=50				

Number of moles in 100 g of H_2 = mass/M_r = 100/2 = 50

According to the reaction coefficients, 2 moles of H_2 produce 2 moles of H_2O

Therefore, 50 moles of H_2 will produce 50 moles of H_2O

Mass of 50 moles of H_2O = number of moles x M_r = 50 x 18 = **900g**

c) What mass of hydrogen can be combusted by 4 g of oxygen gas ?

Answer:

	$2H_{2(g)}$	+	$O_{2(g)}$	$\rightarrow$	$CO_{2(g)}$
A_r/M_r	2		32		18
Mass Balance	4		32		2x18=36
		36			
Reaction Coefficients	2		1		2
Mass(g)	0.25 x 2 = **0.5**		4		
Number of Moles	2 x 0.125 = 0.25		4/32 = 0.125		

Number of moles in 4 g of O_2 = mass/M_r = 4/32 = 0.125
According to the reaction coefficients, 1 mole of O_2 can combust 2 moles of H_2
Therefore 0.125 moles of O_2 can combust 2 x 0.125 = 0.25 moles of H_2
Mass of 0.25 moles H_2 = mass x M_r = 0.25 x 2 = **0.5g**

Example 3.3: *Using the following balanced chemical equation to answer following questions:*

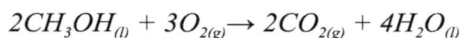

$$2CH_3OH_{(l)} + 3O_{2(g)} \rightarrow 2CO_{2(g)} + 4H_2O_{(l)}$$

a) How many moles of water will be produced from the combustion of 0.17 moles of CH_3OH ?

Answer:

	$2CH_3OH_{(l)}$	+	$3O_{2(g)}$	$\rightarrow$	$2CO_{2(g)}$	+	$4H_2O_{(l)}$
M_r	32		32		44		18
Mass Balance	2x32=64		3x16=96		2x44=88		4x18=72
		160				160	
Reaction Coefficients	2		3		2		4

The reaction coefficients tell us that 2 moles of CH_3OH on complete combustion produces 4 moles of H_2O
Thus 1 mole of CH_3OH produces 2 moles of H_2O
Therefore, 0.17 moles CH_3OH will produce 2 x 0.17 = **0.34 moles of H_2O**

b) How many moles of O_2 are needed to burn 2.78 moles of CH_3OH ?

Answer:

	$2CH_3OH_{(l)}$	+	$3O_{2(g)}$	→	$2CO_{2(g)}$	+	$4H_2O_{(l)}$
M_r	32		32		44		18
Mass Balance	64		96		88		72
		160				160	
Reaction Coefficients	2		3		2		4
Number of Moles	2.78		3/2 x 2.78 = **4.17**				

The reaction coefficients tell us that 2 moles of CH_3OH require 3 moles of O_2 for complete combustion

Thus 1 mole of CH_3OH requires 3/2 moles of O_2 for complete combustion

Therefore, 2.78 moles of CH_3OH requires 3/2 x 2.78 = **4.17 moles of O_2** for complete combustion

c) How many moles of CO_2 are produced from the combustion of 1.25 moles of CH_3OH ?

Answer:

	$2CH_3OH_{(l)}$	+	$3O_{2(g)}$	→	$2CO_{2(g)}$	+	$4H_2O_{(l)}$
M_r	32		32		44		18
Mass Balance	64		96		88		72
		160				160	
Reaction Coefficients	2		3		2		4
Number of Moles	1.25		1.875		**1.25**		2.50

The reaction coefficients tell us that 2 moles of CO_2 are produced from the combustion of 2 moles of CH_3OH

Thus 1 mole of CO_2 is produced from the combustion of 1 mole of CH_3OH

Therefore, **1.25 of CO_2** are produced from the combustion of 1.25 moles of CH_3OH

Example 3.4: *Magnesium metal (Mg) burns with a bright white flame to produce magnesium oxide (MgO). How much MgO could be produced from 2 g of Mg ?*

Answer:

Step 1: Write a balanced equation for the reaction and construct a moles calculating frame around it:

$$2Mg_{(s)} + O_{2(g)} \rightarrow 2MgO_{(s)}$$

Fill in the M_r of all species, identify the reaction coefficients (*i.e.* reaction stoichiometry) and undertake a mass balance {reactants mass = (48 + 32) = products mass (80)} to both ensure that you have a balanced chemical equation and check that you have calculated the M_r values correctly.

	$2Mg_{(s)}$	+	$O_{2(s)}$	→	$2MgO_{(s)}$
M_r	24		32		40
Mass Balance	2x24=48		32		2x40=80
		80			80
Reaction Coefficients	2		1		2
Mass (g)					
Number of Moles					

Step 2: Fill in all the information you have been given in the question:

	$2Mg_{(s)}$	+	$O_{2(s)}$	→	$2MgO_{(s)}$
M_r	24		32		40
Mass Balance	48		32		80
					80
Reaction Coefficients	2		1		2
Mass (g)	2				?
Number of Moles					

Step 3: Work out the number of moles of any species for which you are given the mass (*i.e.* Mg - reactant), using number of moles = mass/M_r. Then use the reaction coefficients to determine the number of moles of the unknown, *i.e.* one (1) mole of Mg will produce one (1) mole of MgO:

	$2Mg_{(s)}$	+	$O_{2(s)}$	→	$2MgO_{(s)}$
M_r	24		32		40
Mass Balance	48		32		80
Reaction Coefficients	2		1		2
Mass (g)	2				?
Number of Moles	2/24=0.083				0.083

Step 4: Since we now know the M_r and the number of moles of the unknown (MgO), its mass can be calculated *i.e.* **mass: m = n x M_r**

	$2Mg_{(s)}$	+	$O_{2(s)}$	→	$2MgO_{(s)}$
M_r	24		32		40
Mass Balance	48		32		80
Reaction Coefficients	2		1		2
Mass (g)	2				0.083 x 40 = **3.32**
Number of Moles	0.083				0.083

Remember you can increase or decrease the number of parameters in the mole calculating frame depending on what you are required to calculate.

Answer: **3.32 g**

Example 3.5: *Baking soda (sodium hydrogen carbonate) thermally degrades to produce sodium carbonate, water and carbon dioxide according the following equation:*

$$2NaHCO_{3(s)} → Na_2CO_{3(s)} + H_2O_{(l)} + CO_{2(g)}$$

Calculate the mass of sodium hydrogen carbonate required to produce 10.6 g of sodium carbonate

Step 1: Construct a moles calculating frame around the balanced chemical equation. Fill in M_r values, identify the mole ratios and undertake a mass balance:

	$2NaHCO_{3(s)}$	→	$Na_2CO_{3(s)}$	+	H_2O	$CO_{2(g)}$
M_r	84		106		18	44
Mass Balance	2 x 84 = 168		106		18	44

Table contd.....

	$2NaHCO_{3(s)}$	$\rightarrow$	$Na_2CO_{3(s)}$	+	H_2O	$CO_{2(g)}$
		168		168		
Reaction Coefficients	2		1		1	1
Mass (g)						
Number of Moles						

Step 2: Fill in all the information you have been given in the question:

	$2NaHCO_{3(s)}$	$\rightarrow$	$Na_2CO_{3(s)}$	+	H_2O	$CO_{2(g)}$
M_r	84		106		18	44
Mass Balance	168		106		18	44
		168		168		
Reaction Coefficients	2		1		1	1
Mass (g)	?		10.8			
Number of Moles						

Step 3: Work out the number of moles of any species for which you are given the mass (*i.e.* Na_2CO_3 - product), using n = m/M_r. Then use the reaction coefficients to determine the number of moles of the unknown, *i.e.* two (2) moles of a HCO_3 produce one (1) mole of Na_2CO_3:

	$2NaHCO_{3(s)}$	$\rightarrow$	$Na_2CO_{3(s)}$	+	H_2O	$CO_{2(g)}$
M_r	84		106		18	44
Mass Balance	168		106		18	44
		168		168		
Reaction Coefficients	2		1		1	1
Mass (g)	?		10.8			
Number of Moles	2 x 0.1 = 0.2		10.6/106= 0.1			

Step 4: Since we now know the M_r and the number of moles of the unknown ($NaHCO_3$), its mass can be calculated *i.e.* **mass: m = n x M_r.** Mass of 0.2 moles of $NaHCO_3$ = 0.2 x 84 = 16.8 g.

	$2NaHCO_{3(s)}$	$\rightarrow$	$Na_2CO_{3(s)}$	+	H_2O	$CO_{2(g)}$
M_r	84		106		18	44

Table contd.....

	$2NaHCO_{3(s)}$	$\rightarrow$	$Na_2CO_{3(s)}$	+	H_2O	$CO_{2(g)}$
Mass Balance	168		106		18	44
		168		168		
Reaction Coefficients	2		1		1	1
Mass (g)	0.2 x 84 = **16.8**		10.8			
Number of Moles	0.2		0.1			

Answer: **16.8 g**

Exercise 3.2

a) *Complete combustion of a sample of butane, C_4H_{10} generated 2.46 g of water.*

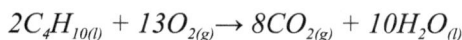

$$2C_4H_{10(l)} + 13O_{2(g)} \rightarrow 8CO_{2(g)} + 10H_2O_{(l)}$$

     i. *Determine the number of moles of water generated ?*
     ii. *How many moles of butane were combusted ?*
     iii. *What mass of butane was combusted ?*
     iv. *How many moles of oxygen were consumed ?*
     v. *Calculate the mass of oxygen used up in grams ?*

b) *Determine the number of moles of N_2O_4 needed to react completely with 3.62 moles of N_2H_4 for the reaction:*

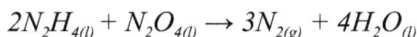

$$2N_2H_{4(l)} + N_2O_{4(l)} \rightarrow 3N_{2(g)} + 4H_2O_{(l)}$$

c) *What mass of iron (II) sulfide can be formed from 14.0 g of sulphur ?*

$$Fe_{(s)} + S_{(s)} \rightarrow FeS_{(s)}$$

d) *Determine the mass of aluminium required to completely react with 10 g of CuO and calculate the mass of Al_2O_3 produced ?*

$$3CuO_{(s)} + 2Al_{(s)} \rightarrow Al_2O_{3(s)} + 3Cu_{(s)}$$

e) *What mass of magnesium oxide is made when 250 g of oxygen reacts with excess magnesium ?*

$$2Mg_{(s)} + O_{2(g)} \rightarrow 2MgO_{(s)}$$

f) Titanium(IV) chloride can be converted to titanium by reacting it with an excess of magnesium.

$$TiCl_{4(l)} + 2Mg_{(s)} \rightarrow Ti_{(s)} + 2MgCl_{2(s)}$$

Determine the mass of titanium that could theoretically be produced from 37.98 kg of titanium(IV) chloride ?

g) Calculate the mass of potassium hydrogen carbonate required to produce 100 g of potassium carbonate on thermal decomposition ?

$$2KHCO_{3(s)} \rightarrow K_2CO_{3(s)} + CO_{2(g)} + H_2O_{(l)}$$

h) The reaction below forms the basis of thermite welding which is often used to join rail tracks.

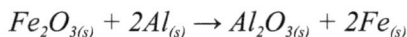

$$Fe_2O_{3(s)} + 2Al_{(s)} \rightarrow Al_2O_{3(s)} + 2Fe_{(s)}$$

What mass of aluminium is required to produce 7g of iron ?

i) Determine the mass of zinc chloride that could be produced from 10 g of zinc.

$$Zn_{(s)} + 2HCl_{(aq)} \rightarrow ZnCl_{2(aq)} + H_{2(g)}$$

j) Lithium oxide is used as a drying on the space shuttle. What mass of water could be removed by 65g of lithium oxide ?

$$Li_2O_{(s)} + H_2O_{(l)} \rightarrow 2LiOH_{(s)}$$

k) Cisplatin is an anti-cancer agent prepared as follows:

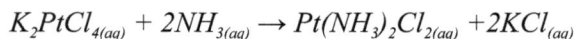

$$K_2PtCl_{4(aq)} + 2NH_{3(aq)} \rightarrow Pt(NH_3)_2Cl_{2(aq)} + 2KCl_{(aq)}$$

Calculate the mass of cisplatin that could be generated from 10.0 g of K_2PtCl_4 ?

l) The combustion of ammonia (NH_3) produces nitrogen dioxide:

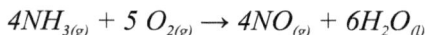

$$4NH_{3(g)} + 5\,O_{2(g)} \rightarrow 4NO_{(g)} + 6H_2O_{(l)}$$

Determine the number of moles and the mass of oxygen (O_2) required to react with 56.8 grams of ammonia ?

m) How many grams of sodium sulfate could be produced from 200 g of sodium hydroxide ?

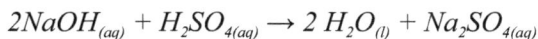

$$2NaOH_{(aq)} + H_2SO_{4(aq)} \rightarrow 2\ H_2O_{(l)} + Na_2SO_{4(aq)}$$

n) Using the following equation, calculate the number of moles and mass of iodine (I_2) that could be generated from 8.2 grams of $NaIO_3$.

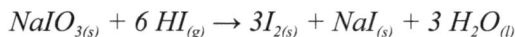

$$NaIO_{3(s)} + 6\ HI_{(g)} \rightarrow 3I_{2(s)} + NaI_{(s)} + 3\ H_2O_{(l)}$$

o) Copper hydroxide is precipitated when a solution of copper sulphate reacts with a solution of sodium hydroxide, according to the following equation:

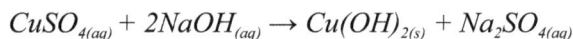

$$CuSO_{4(aq)} + 2NaOH_{(aq)} \rightarrow Cu(OH)_{2(s)} + Na_2SO_{4(aq)}$$

Calculate the mass of sodium hydroxide required to transform 15.95 g of copper sulphate to copper hydroxide and the mass of copper hydroxide that would be produced.

ANSWERS

Exercise 3.1

Balance the following chemical equations

a. $2Mg + O_2 \rightarrow 2MgO$
b. $Ca + 2H_2O \rightarrow Ca(OH)_2 + H_2$
c. $CuCO_3 + H_2SO_4 \rightarrow CuSO_4 + H_2O + CO_2$
d. $CH_4 + 2H_2O \rightarrow CO_2 + 4H_2$
e. $3NO_2 + H_2O \rightarrow 2HNO_3 + NO$
f. $2NaCl + 2H_2O \rightarrow 2NaOH + Cl_2 + H_2$
g. $CaCl_2 + Na_2CO_3 \rightarrow CaCO_3 + 2NaCl$
h. $C_{12}H_{22}O_{11} + 12O_2 \rightarrow 12CO_2 + 11H_2O$
i. $(NH_4)_2CO_3 \rightarrow 2NH_3 + CO_2 + H_2O$
j. $2Fe + 3Cl_2 \rightarrow 2FeCl_3$
k. $4Fe + 3O_2 \rightarrow 2Fe_2O_3$
l. $3Ba(OH)_2 + 2\ H_3PO_4 \rightarrow 6\ H_2O + Ba_3(PO_4)_2$
m. $2N_2O_5 \rightarrow 4\ NO_2 + O_2$

n. $CaCl_2 + 2\,AgNO_3 \rightarrow 2\,AgCl + Ca(NO_3)_2$

o. $Na_2SO_4 + 2C + CaCO_3 \rightarrow Na_2CO_3 + CaS + 2CO_2$

And finally:

$$4\,CuSCN + 7KIO_3 + 14HCl \rightarrow 4\,CuSO_4 + 7KCl + 4HCN + 7ICl + 5H_2O$$

Exercise 3.2

Answers

a) *Complete combustion of a sample of butane, C_4H_{10} generated 2.46 g of water.*

$$2C_4H_{10(l)} + 13O_{2(g)} \rightarrow 8CO_{2(g)} + 10H_2O_{(l)}$$

    i. *Determine the number of moles of water generated ?*

    ii. *How many moles of butane were combusted ?*

    iii. *What mass of butane was combusted ?*

    iv. *How many moles of oxygen were consumed ?*

    v. *Calculate the mass of oxygen used up in grams ?*

Answer:

	$2C_4H_{10(l)}$	+	$13O_{2(g)}$	$\rightarrow$	$8CO_{2(g)}$	+	$10H_2O_{(l)}$
A_r/M_r	58		32		44		18
Mass Balance	58 x 2 = 116		32 x 13 = 416		8 x 44 = 352		18 x 10 = 180
		532				532	
Reaction coefficients	2		13		8		10
Mass (g)	**1.6**		0.179 x 32 = **5.74**		4.8		2.46
No. of moles	1.6/58 = **0.0274**		0.179		0.1093		2.46/18 = **0.137**

    i. Number of moles of in 2.46 g of H_2O = mass/M_r = 2.46/18 = **0.137** moles

    ii. According to the reaction coefficients, the combustion of 1 mole of C_4H_{10} generates 5 moles of H_2O

       Number of moles of C_4H_{10} combusted to produce 0.137 moles of H_2O = 1/5 x 0.137 = **0.0274**

    iii. Mass of 0.0274 moles of C_4H_{10} = number of moles x M_r = 0.0274 x

$58 = \textbf{1.6 g}$

iv. According to the reaction coefficients, 10 moles of H_2O are generated from 13 moles of O_2

Thus the number of moles of O_2 consumed to produced 0.137 moles of H_2O = 13/10 x 0.137 = **0.179 moles of O_2**

v. Mass of 0.179 moles of O_2 = number of moles x M_r = 0.179 x 32 = **5.74 g**

b) Determine the number of moles of N_2O_4 needed to react completely with 3.62 moles of N_2H_4 for the reaction:

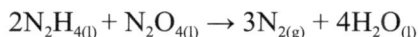

$$2N_2H_{4(l)} + N_2O_{4(l)} \rightarrow 3N_{2(g)} + 4H_2O_{(l)}$$

Answer:

The balanced chemical equation tells us that 2 moles of $2N_2H_4$ react with 1 mole of N_2O_4 to produce 3 moles of N_2 and 4 moles of H_2O

The number of moles of N_2O_4 required to react completely with 3.62 moles of N_2H_4 = 3.62/2 = **1.81 moles of N_2O_4**

c) What mass of iron (II) sulfide can be formed from 14.0 g of sulphur ?

$$Fe_{(s)} + S_{(s)} \rightarrow FeS_{(s)}$$

Answer:

	$Fe_{(s)}$	+	$S_{(s)}$	$\rightarrow$	$FeS_{(s)}$
A_r/M_r	56		32		88
Mass Balance		88			88
Reaction Coefficients	1		1		1
Mass (g)	0.438 x 56 = 24.5		14		0.438 x 88 =**38.5**
No. of moles	0.438		14/32 = 0.438		

Number of moles in 14g of S = 14/32 = 0.438

According to the reaction coefficients, 1 mole of Fe reacts with 1 mole of S to form 1 mole of FeS

Therefore, 0.438 moles of S will produce 0.438 moles of FeS

Mass of 0.438 moles of FeS = number of moles x M_r = 0.438 x 88 = **38.5g**

d) Determine the mass of aluminium required to completely react with 10 g of CuO and calculate the mass of Al_2O_3 produced ?

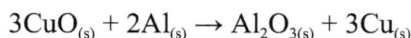

$$3CuO_{(s)} + 2Al_{(s)} \rightarrow Al_2O_{3(s)} + 3Cu_{(s)}$$

Answer:

	3CuO$_{(s)}$	+	**2Al**	→	**Al$_2$O$_{3(s)}$**	+	**3Cu**$_{(s)}$
A$_r$/ M$_r$	79.5		27		102		63.5
Mass Balance	3 x 79.5 = 238.5		54		102		3 x 63.5 = 190.5
		292.5				292.5	
Reaction Coefficients	3		2		1		3
Mass (g)	10				0.042 x 102 = **4.28 g**		
No. of moles	10/79.5 = 0.126				0.126/3 = 0.042		

Number of moles in 10 g of CuO = mass/ M_r = 10/79.5 = 0.126

According to the reaction coefficients, 3 moles of CuO produce 1 mole of Al2O3

Therefore, the number of moles of Al2O3 produced from 0.126 moles of CuO = 0.126/3 = 0.042

Mass of 0.042 moles of Al2O3 = number of moles x M_r = 0.042 x 102 = **4.28 g**

e) What mass of magnesium oxide is made when 250 g of oxygen reacts with excess magnesium ?

$$2Mg_{(s)} + O_{2(g)} \rightarrow 2MgO_{(s)}$$

Answer:

	2Mg$_{(s)}$	+	**O**$_{2(g)}$	→	**2MgO**$_{(s)}$
A$_r$/M$_r$	24		32		40
Mass Balance	48		32		2 x 40 = 80
Mass Balance		80			80
Reaction Coefficients	2		1		2
Mass (g)			250		15.6 x 40 = **624 g**

Table contd.....

	$2Mg_{(s)}$	+	$O_{2(g)}$	$\rightarrow$	$2MgO_{(s)}$
No. of moles			$250/32 = 7.8$		$2 \times 7.8 = 15.6$

Number of moles in 250 g of O_2 = $250/32 = 7.8$

According to the reaction coefficients, one mole of O_2 produces 2 moles of MgO

Therefore, the number of moles of MgO produced from 7.8 moles of O2 = 7.8 x 2 = 15.6

Mass of 15.6 moles of MgO = number of moles x M_r = 15.6 x 40 = **624 g**

f) Titanium(IV) chloride can be converted to titanium by reacting it with an excess of magnesium.

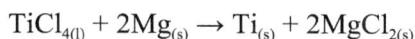

$$TiCl_{4(l)} + 2Mg_{(s)} \rightarrow Ti_{(s)} + 2MgCl_{2(s)}$$

What mass of titanium could theoretically be obtained from 37.98 kg of titanium(IV) chloride ?

Answer:

	$TiCl_{4(l)}$	+	$2Mg_{(s)}$	$\rightarrow$	$Ti_{(s)}$	+	$2MgCl_{2(s)}$
A_r/ M_r	190		24		48		95
Mass Balance	190		$2 \times 24 = 48$		48		$2 \times 95 = 190$
		138				138	
Reaction Coefficients	1		2		1		2
Mass (g)	37980				$199.9 \times 48 = $ **9595**		
No. of moles	$37980/190 = 199.9$				200		

Number of moles in 37.49 kg (37,490 g) of $TiCl_4$ = $37980/190 = 199.9$

According to the reaction coefficients, 1 mole of $TiCl_4$ generates 1 mole of Ti

g) Calculate the mass of potassium hydrogen carbonate required to produce 100 g of potassium carbonate on thermal decomposition ?

$$2KHCO_{3(s)} \rightarrow K_2CO_{3(s)} + CO_{2(g)} + H_2O_{(l)}$$

Answer:

	2KHCO$_{3(aq)}$	→	K$_2$CO$_{3(aq)}$	+	H$_2$O$_{(aq)}$	+	CO$_{2(aq)}$
M$_r$	100				18		44
Mass Balance	200		138		18		44
		200		200			
Reaction Coefficients	2		1		1		1
Mass (g)	1.45 x 100 = **145**		100				
No. of moles	0.725 x 2 = 1.45		100/138 = 0.725				

Number of moles of in 100 g of K$_2$CO$_3$ = 100/138 = 0.725
According to the reaction coefficients, 1 mole of K$_2$CO$_3$ is produced from 2 moles of KHCO$_3$
Therefore, 0.725 moles of K$_2$CO$_3$ would be generated from 0.725 x2 = 1.45 moles of KHCO$_3$
Mass of 1.45 moles of KHCO$_3$ = number of moles x M$_r$ = 1.45 x 100 = **145 g**

h) The reaction below forms the basis of thermite welding which is often used to join rail tracks.

What mass of aluminium is required to produce 7g of iron ?

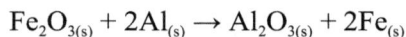

$$Fe_2O_{3(s)} + 2Al_{(s)} \rightarrow Al_2O_{3(s)} + 2Fe_{(s)}$$

Answer:

	Fe$_2$O$_{3(s)}$	+	2Al$_{(s)}$	→	Al$_2$O$_{3(s)}$	+	2Fe$_{(s)}$
	160		27		102		56
Mass Balance	160		54		102		112
		214				214	
Reaction Coefficients	1	:	2	:	1	:	2
Mass (g)			0.125 x 27 = **3.375**				7
No. of moles			0.125				7/56 = 0.125

Number of moles in 7g of Fe = 7/56 = 0.125
According to the reaction coefficients, 2 moles of Fe are generated from 2 moles of Al

Therefore, the number of moles of Al required to produce 0.125 moles of Fe = 0.125

Mass of 0.125 moles of Al = number of moles x M_r = 0.125 x 27 = **3.375 g**

i) Determine the mass of zinc chloride that could be produced from 10 g of zinc.

$$Zn_{(s)} + 2HCl_{(aq)} \rightarrow ZnCl_{2(aq)} + H_{2(g)}$$

Answer:

	$Zn_{(s)}$	+	$2HCl_{(aq)}$	$\rightarrow$	$ZnCl_{2(aq)}$	+	$H_{2(g)}$
A_r / M_r	65		36.5		136		2
Mass Balance	65		73		136		2
		138				138	
Reaction Coefficients	1		2		1		1
Mass (g)	10				0.154 x 136 = **20.9**		
No. of moles	10/65 = 0.154				0.154		

Number of moles in 10 g of Zn = 10/65 = 0.154

According to the reaction coefficients, 1 mole of Zn produces 1 mole of $ZnCl_2$

Therefore, the number of moles of $ZnCl_2$ generated from 0.154 moles of Zn = 0.154

Mass of 0.154 moles of $ZnCl_2$ = number of moles x M_r = 0.154 x 136 = **20.9 g**

j) Lithium oxide is a drying agent used on the space shuttle. What mass of water could be removed by 65g of lithium oxide ?

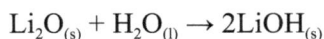

$$Li_2O_{(s)} + H_2O_{(l)} \rightarrow 2LiOH_{(s)}$$

Answer:

	$Li_2O_{(s)}$	+	$H_2O_{(l)}$	$\rightarrow$	$2LiOH_{(s)}$
A_r / M_r	30		18		24
Mass Balance	30	48	18		48
Reaction Coefficients	1	:	1	:	2
Mass (g)	65		2.17 x 18 = **39**		
No. of moles	65/30 = 2.17		2.17		

Number of moles in 65 g of Li_2O = 65/30 = 2.17

According to the reaction coefficients, 1 mole of Li_2O reacts with 1 mole of H_2

Therefore 2.17 moles Li_2O will react with 2.17 moles of H_2O

Mass of 2.17 moles of H_2O = number of moles x M_r = 2.17 x 18 = 39 g

k) Cisplatin is an anti-cancer agent prepared as follows:

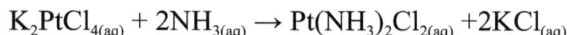

$$K_2PtCl_{4(aq)} + 2NH_{3(aq)} \rightarrow Pt(NH_3)_2Cl_{2(aq)} + 2KCl_{(aq)}$$

Calculate the mass of cisplatin that could be generated from 10.0 g of K_2PtCl_4 ?

Answer:

	$K_2PtCl_{4(aq)}$	+	$2NH_{3(aq)}$	$\rightarrow$	$Pt(NH_3)_2Cl_{2(aq)}$	+	$2KCl_{(aq)}$
A_r/ M_r	415		17		300		74.5
Mass Balance	415		34		300		149
		449				449	
Reaction Coefficients	1	:	2	:	1	:	2
Mass (g)	10				7.24		
No. of moles	10/417 = 0.024				0.024		

Number of moles in 10 g of K_2PtCl_4 = 10/415 = 0.024

According to the equation, 1 mole of K_2PtCl_4 generates 1 mole of $Pt(NH_3)_2C_{12}$

Therefore, 0.024 moles of K_2PtCl_4 generates 0.024 moles of $Pt(NH_3)_2C_{12}$

Mass of 0.024 moles of $Pt(NH_3)_2C_{12}$ = number of moles x M_r = 0.024 x 302 = **7.24 g**

l) The combustion of ammonia (NH_3) produces nitrogen dioxide:

$$4NH_{3(g)} + 5\ O_{2(g)} \rightarrow 4NO_{(g)} + 6H_2O_{(l)}$$

Determine the number of moles and the mass of oxygen (O_2) required to react with 56.8 grams of ammonia ?

Answer:

	$4NH_{3(g)}$	+		$5O_{2(g)}$		→	$4NO_{(g)}$	+	$6H_2O_{(l)}$
A_r/ M	17			32			30		18
Mass Balance	68			160			120		108
		228						228	
Reaction Coefficients	4			5			4		6
Mass (g)	56.8			4.18 x 32 = **133.6**					
No. of moles	56.8/17 = 3.34			5/4 x 3.34 = 4.176					

Number of moles in 56.8 g of NH_3 = 56.8/17 = 3.34

According to the reaction coefficients 4 moles of NH_3 requires 5 moles of O_2

Thus each mole of NH_3 requires 5/4 moles of O_2

Therefore, 3.34 moles of NH_3 requires 5/4 x 3.34 moles = 4.18 moles O_2

Mass of 4.18 moles O_2 = number of moles x M_r= 4.18 x 32 = **133.6 g**

m) How many grams of sodium sulfate could be produced from 200 g of sodium hydroxide ?

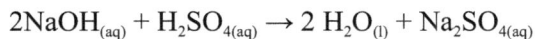

$$2NaOH_{(aq)} + H_2SO_{4(aq)} \rightarrow 2\ H_2O_{(l)} + Na_2SO_{4(aq)}$$

Answer:

	$2NaOH_{(aq)}$	+	$H_2SO_{4(aq)}$	→	$2H_2O_{(l)}$	+	$Na_2SO_{4(aq)}$
A_r/ M_r	40		98		18		142
Mass Balance	80		98		36		142
		178				178	
Reaction coefficients	2		1		2		1
Mass	200						2.5 x 142 = **355**
No. of moles	200/40=5		2.5		5		2.5

Number of moles in 200g of NaOH = mass/M_r= 200/40 = 5

According to the reaction coefficients, 2 moles of NaOH produce 1 mole Na_2SO_4

Therefore, 5 moles of NaOH will produce 5 x 0.5 = 2.5 moles of Na_2SO_4

Mass of 2.5 moles of Na_2SO_4 = number of moles x M_r= 2.5 x 141 = **355 g**

n) Using the following equation, calculate the number of moles and mass of iodine (I_2) that could be generated from 8.2 grams of $NaIO_3$.

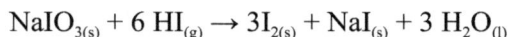

$$NaIO_{3(s)} + 6\ HI_{(g)} \rightarrow 3I_{2(s)} + NaI_{(s)} + 3\ H_2O_{(l)}$$

Answer:

	NaIO$_3$	+	6HI	→	3I$_2$	+	4NO	+	3H$_2$O
A$_r$/ M$_r$	198		128		254		30		18
Mass Balance	198		768		762		120		54
		966				966			
Reaction coefficients	1		6		3		4		3
Mass	8.2				0.124 x 254 = **31.5**				
No. of moles	8.2/198 = 0.0414				0.124				

Number of moles in 8.2 g of $NaIO_3$ = mass/M_r = 8.2/198 = 0.0414
According to the reaction coefficients, 1 mole of $NaIO_3$ produces 3 moles I_2
Therefore, 0.0414 moles of $NaIO_3$ will produce 3 x 0.0414 = 0.124 moles of I_2
Mass of 0.124 moles of I_2 = number of moles x M_r = 0.124 x 254 = **31.6 g**

o) Copper hydroxide is precipitated when a solution of copper sulphate reacts with a solution of sodium hydroxide, according to the following equation:

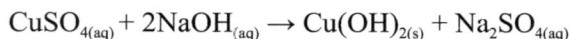

$$CuSO_{4(aq)} + 2NaOH_{(aq)} \rightarrow Cu(OH)_{2(s)} + Na_2SO_{4(aq)}$$

Calculate the mass of sodium hydroxide required to transform 15.95 g of copper sulphate to copper hydroxide and the mass of copper hydroxide that would be produced.

Answer:

	CuSO$_{4(aq)}$	+	2NaOH$_{(aq)}$	→	Cu(OH)$_{2(aq)}$	+	Na$_2$SO$_{4(aq)}$
A$_r$/ M$_r$	159.5		40		97.5		142
Mass Balance	159.5		80		97.5		142
		239.5				239.5	
Reaction coefficients	1		2		1		1
Mass (g)	15.95		0.2 x 40 = **8**		0.1 x 97.5 = **9.75**		14.2
No. of moles	159.5/159.5 = 0.1		0.2		0.1		0.1

Number of moles in 15.95 g of $CuSO_4$ = mass/M_r = 15.95/159.5 = 0.1

According to the reaction coefficients, 1 mole of $CuSO_4$ reacts with 2 moles of $NaOH$ to produce 1 mole of $Cu(OH)_2$

Therefore, 0.1 moles $CuSO_4$ reacts with 0.2 moles of $NaOH$ to produce 0.1 mole of $Cu(OH)_2$

Mass of 0.2 moles $NaOH$ = number of moles x M_r = 0.2 x 40 = **8 g**

Mass of 0.1 moles $Cu(OH)_2$ = number of moles x M_r = 0.1 x 97.5 = **9.75 g**

CHAPTER 4

Molarity & Concentration

Keywords: Calculations, Concentration, Molarity, Mole fraction, Parts per million, Percent by mass, Percent by volume.

4.1. CONCENTRATION

Most chemical processes occur in solution. "Life" has been described as the sum of a series of complex processes occurring in solution. Air, seawater, tea, beer and toilet bleach are all solutions. A solution is simply a homogenous mixture of substances of variable composition. The most abundant substance is called the **solvent**, whereas the substance present in lesser amounts is called the **solute**. If a small quantity of ethanol is added to water, the ethanol is the solute and the water is the solvent. But if we add a smaller amount of water to a larger amount of ethanol, then the water is the solute and the ethanol is the solvent. Although we will predominantly be concerned with solutions produced by solids dissolved in liquids, there are as many types of solutions as there are different combinations of solids, liquids, and gases. Brass is solid solution of copper and zinc, whilst the atmosphere is a solution in which a gaseous solvent (nitrogen) dissolves other gases (such as oxygen, carbon dioxide, water vapour, and neon).

Scientists often refer to concentrated solutions, dilute solutions, or very dilute solutions. Dilute solutions contain a relatively small amount of the solute in a given volume of solvent. Tap water is an example of a dilute solution since it contains small quantities of dissolved minerals. A concentrated solution on the other hand has a large amount of solute in the solvent.

Concentration can be expressed in several ways:

Nigel P. Freestone
All rights reserved-© 2016 Bentham Science Publishers

Solution Type (solute - solvent)	Concentration Units	Concentration Equation
Solid-Liquid	Molarity (M or mol/L or mol dm^{-3})	Number of moles of solute per litre of solvent *
Solid-Liquid	Mass per volume (g/L or g dm^{-3})	Number of g per litre of solvent
Solid-Solid	Percent by Mass (m/m %) *	Mass of solute/ Mass of solution x 100
Liquid-Liquid	Percent by volume (v/v %)	Volume of solute / Volume of Solution x 100
Solid-Liquid	Percent by mass/volume (m/v %) **	Mass of Solute/ Volume of Solution x 100
All Solutions	Parts per million (ppm or mg /L or mg/kg)	Mass of solute/ Mass of Solution x 10^6

IMPORTANT NOTE:

* m/m % is often improperly referred to as weight percent (wt %) or weight-weight percent (w/w %)

** m/v % is often improperly referred to as % weight-volume percent (w/v %)

Since mass and weight are different quantities, both w/w % and w/v % are both incorrect.

4.2. MOLARITY

The concentration of a solution is expressed in terms of the amount of solute present in a standard volume of solvent. The standard volume is 1 litre, which is the same as is the same as 1 cubic decimetre (1 dm^3), which is the same as 1,000 cm^3. The most commonly used unit of concentration is molarity, often abbreviated as M. Molarity is simply the concentration of a solution expressed as the number of moles of solute per litre of solution.

A 1 M solution contains a molar mass (M$_r$) of solute in 1 litre of the solvent.

The following units of concentration are all the same:

$$M = mol/L = mol \ dm^{-3}$$

Concentration (c) = Amount of Solute (n) / Volume of solution (v)

c = n/v

Amount of Solute (n) = Volume of Solution (v) x Concentration (c)

n = v x c

Volume of Solution (v) = Concentration (c) / Amount of Solute (n)

v = c/n

Concentration can also be expressed as stated above in terms of mass (g) of solute per unit of volume (litre), *i.e.* g/L.

Units:

V = Volume in litres (L)
Note: to convert cm^3 to L divide by 1000, *e.g* 10 cm^3 = 10/1000 = 0.01 L

If the amount of solute is measured in moles, then concentration unit is M
If amount of solute is measured in g then concentration unit is g/L or g dm^{-3}

For example, glucose, $C_6H_{12}O_6$ has a molar mass (M_r) of 180 g mol^{-1}. Therefore, a solution of 180 g of glucose dissolved in a total volume of 1 litre (L) has a concentration of 1.0 M or 180 g/L. A 0.1 M solution of glucose therefore must contain 18.0 g (0.1 x M_r) of dissolved glucose in 1 L of solution. Similarly 1.8 g of glucose in 100 cm^3 is equivalent to 18 g in 1000 cm^3 (1 L), giving a concentration of 0.1 M or 18 g/L.

To convert from M to g/L simply multiply by M_r
To convert g/L to M simply divide by M_r

Example 4.1: *What is the molarity of an aqueous solution of sodium chloride (NaCl) with a concentration of 5.85 g/L?*

Answer:

Molar Mass (M_r) of NaCl = 58.5
To convert g/L to M divide by M_r
Therefore, Molarity = 5.85/50.5 = **0.1 M**

Example 4.2: *A solution of $KMnO_4$ has a concentration of 0.2M. Express this concentration in terms of g/L.*

Answer:

Molar Mass (M_r) of $KMnO_4$ = 158 g mol^{-1}
To convert M to g/L multiply by M_r
Thus concentration = 0.2 x 158 = **31.6 g/L**

Example 4.3: *What is the concentration of a solution containing 2.016 g of dissolved sodium carbonate (Na_2CO_3) in 100 cm^3 (0.1L) of water?*

Answer:

Molar mass (M_r) of sodium carbonate (Na_2CO_3) = 106 g mol^{-1}
Number of moles of Na_2CO_3 in 2.016 g = 2.016/106 = 0.0190
Concentration = number of moles / volume = 0.0190/0.1 = **0.19 M**

Example 4.4: *How many moles of hydrochloric acid are present in 25 cm^3 of 0.025M hydrochloric acid solution?*

Answer:

Concentration = 0.025 M
Volume = 25 cm^3 = 25/1000 = 0.025 L
Number of moles = concentration x volume = 0.025 x 0.025 = **6.25 x 10^{-4} moles**

Example 4.5: *What mass of sodium hydroxide is present in 200 cm^3 of a 0.01 M sodium hydroxide solution?*

Answer:

Concentration = 0.01M
Volume, V = 200 cm^3 = 200/1000 = 0.2 L
Number of moles = concentration x volume = 0.01 x 0.2 = 2 x 10^{-3} moles
Molar mass (M_r) of NaOH = 40 g mol^{-1}
Mass of 2 x 10^{-3} moles of NaOH = number of moles x M_r = 2 x 10^{-3} x 40 = **0.08 g**

Example 4.6: Complete the following table:

Mass of glucose ($C_6H_{12}O_6$)	Volume of water (cm^3)	Concentration (g /L)	Concentration (M)
1.8	100	18	0.1
36	250	a)	b)
60	c)	d)	0.33
90	1000	90	e)
180	1000	180	1.0

Answer:

$M_r [C_6H_{12}O_6] = 180$ g mol^{-1}

a. Volume: 250 cm^3 = 250/1000 = 0.25 litres
   Amount of solute, n = 36 g
   Concentration = amount of solute/ volume = 36/0.25 = **144 g/L**
b. Number of moles of $C_6H_{12}O_6$ in 36 g = 36/180 = 0.2 moles
   Thus amount of solute, n = 0.2 moles
   Concentration = amount of solute / volume = 0.2 moles/ 0.25 litres = **0.8M**
c. Concentration = 0.33M
   Number of moles of $C_6H_{12}O_6$ in 60 g = 60/180 = 0.33 moles
   Thus n = 0.33 moles
   Volume = amount of solute /concentration = 0.33/0.33 = **1 L**
d. Concentration = 0.33 M
   To convert from M to g/L multiply by M_r
   Concentration = 0.33 x 180 = **60 g/L**
e. Concentration = 90 g/L
   To convert from g/L to M, divide by M_r = 90/180 = **0.5 M**

Exercise 4.1

Calculate the concentration of the following solutions containing:

a. *2.65 g of dissolved sodium chloride (NaCl) in 200 cm^3 of water.*
b. *10.6 g of dissolved sodium carbonate (Na_2CO_3) in 500 cm^3 of water.*
c. *0.005 g of dissolved oxygen (O_2) in 100 cm^3 of water.*
d. *6 g of dissolved calcium hydroxide {$Ca(OH)_2$} in 800 cm^3 of water.*
e. *0.34 g of dissolved sodium hypochlorite (NaClO) in 50 cm^3 of water.*
f. *1.2 g of dissolved sodium thiosulfate ($Na_2S_2O_3$) in 150 cm^3 of water.*
g. *2.356 g of dissolved in ammonium hydrogen carbonate {$(NH_4)HCO_3$} in 350 cm^3 of water.*
h. *0.3 g of dissolved calcium oxide (CaO) in 250 cm^3 of water.*
i. *0.45 g of dissolved oxalic acid ($C_2H_4O_2$) in 200 cm^3 of water.*
j. *1.24 g of dissolved sodium hydrogen carbonate ($NaHCO_3$) in 75 cm^3 of water.*

Exercise 4.2

What mass of solute must be dissolved in water to make:

a. *100 cm³ of 1 M sodium hydroxide (NaOH) ?*
b. *500 cm³ of 2 M nitric acid (HNO₃) ?*
c. *1 cm³ of 0.05 M silver nitrate (AgNO₃) ?*
d. *25 cm³ of 1.5 x 10⁻³ M sulfuric acid (H₂SO₄) ?*
e. *17.4 cm³ of 0.125 M oxalic acid (C₂H₄O₂) ?*
f. *150 cm³ of 0.12 M sucrose (C₁₂H₂₂O₁₁) ?*
g. *23.5 cm³ of 0.00125 M barium sulfate (BaSO₄) ?*
h. *12.4 cm³ of 0.101 M hydrochloric acid (HCl) ?*
i. *5 cm³ of 2 x 10⁻⁴ M ascorbic acid (C₆H₈O₆) ?*
j. *12.3 cm³ of 1.03 x 10⁻⁴ M iron (III) chloride (FeCl₃)*

Exercise 4.3

How many moles are present in the following solutions:

a. *25 cm³ of 1 M NaOH ?*
b. *7.5 litres of 5 M H₂SO₄ ?*
c. *50 cm³ of 0.101 M NaCl ?*
d. *12.7 cm³ of 0.035 M K₂Cr₂O₇ ?*
e. *2.6 cm³ of 0.102 M H₂O₂ ?*
f. *32.8 cm³ of 0.033 M HNO₃ ?*
g. *12.8 cm³ of 0.65 M NaMnO₄ ?*
h. *42.13 cm³ of 0.05 M AgNO₃ ?*
i. *5.6 litres of 1.35 M NaCl ?*
j. *23.9 cm³ of 1.56 x 10⁻³ M KI ?*

4.3. PERCENTAGE BY MASS (M/M %) AND MASS FRACTION (M/M)

Concentration can also be expressed in terms of a fraction or percent by mass and volume.

Exercise 4.2

What mass of solute must be dissolved in water to make:

a. *100 cm³ of 1 M sodium hydroxide ($NaOH$) ?*
b. *500 cm³ of 2 M nitric acid (HNO_3) ?*
c. *1 cm³ of 0.05 M silver nitrate ($AgNO_3$) ?*
d. *25 cm³ of 1.5×10^{-3} M sulfuric acid (H_2SO_4) ?*
e. *17.4 cm³ of 0.125 M oxalic acid ($C_2H_4O_2$) ?*
f. *150 cm³ of 0.12 M sucrose ($C_{12}H_{22}O_{11}$) ?*
g. *23.5 cm³ of 0.00125 M barium sulfate ($BaSO_4$) ?*
h. *12.4 cm³ of 0.101 M hydrochloric acid (HCl) ?*
i. *5 cm³ of 2×10^{-4} M ascorbic acid ($C_6H_8O_6$) ?*
j. *12.3 cm³ of 1.03×10^{-4} M iron (III) chloride ($FeCl_3$)*

Exercise 4.3

How many moles are present in the following solutions:

a. *25 cm³ of 1 M $NaOH$?*
b. *7.5 litres of 5 M H_2SO_4 ?*
c. *50 cm³ of 0.101 M $NaCl$?*
d. *12.7 cm³ of 0.035 M $K_2Cr_2O_7$?*
e. *2.6 cm³ of 0.102 M H_2O_2 ?*
f. *32.8 cm³ of 0.033 M HNO_3 ?*
g. *12.8 cm³ of 0.65 M $NaMnO_4$?*
h. *42.13 cm³ of 0.05 M $AgNO_3$?*
i. *5.6 litres of 1.35 M $NaCl$?*
j. *23.9 cm³ of 1.56×10^{-3} M KI ?*

4.3. PERCENTAGE BY MASS (M/M %) AND MASS FRACTION (M/M)

Concentration can also be expressed in terms of a fraction or percent by mass and volume.

Mass Fraction

Mass fraction (m/m) is simply the ratio of the mass of one component of a solution to the total mass of the solution.

Mass fraction of Component A = Mass of A / Total Mass of Solution
m/m % of A = Mass of Component A / Total Mass of Solution x 100

Example 4.7: *A wedding ring with a mass of 15.0 grams contains 14.1 grams of pure silver. What is the concentration of silver (m/m %)?*

Answer:

Concentration (m/m %) = mass of component/ total mass x 100 = 14.1/15 x 100 = **94 %**

Note: m/m % = w/w % = wt%

In examples and exercises both w/w % and m/m % concentration units will be used.

Example 4.8: *A 10.4 g aqueous solution contains 450 mg of sodium chloride. What is the mass fraction of sodium chloride?*

Answer:

Convert all masses to the same units:
Mass (NaCl) - 450 mg = 450 x 10^{-3} = 0.45 g
Mass fraction (NaCl) = Mass of NaCl/ Total mass of solution = 0.45 / 10.4 = **0.43**

Example 4.9: *If 200 g of an aqueous solution of fructose contains a mole fraction of water of 0.65, what mass of sucrose is present?*

Answer:

Mass fraction of fructose	= 1 – mass faction of water
	= 1 – 0.65 = 0.35
Mass of fructose	= mass fraction of fructose x total solution mass

$$= 0.35 \times 200 = \textbf{70 g}$$

Example 4.10: *An aqueous solution of sodium nitrate has a mass % water of 68%. What is the mass percentage of sodium nitrate?*

Answer:

% Mass (KCl) + Mass % (H_2O) = 100

% Mass (KCl) = 100 − Mass % (H_2O)

$\qquad\qquad\quad$ = 100 − 68

$\qquad\qquad\quad$ = **32%**

Example 4.11: *What mass of ethanol is present in 500 g of an aqueous solution containing 35% ethanol by mass?*

Answer:

Mass fraction of ethanol = Mass of ethanol / Total mass of solution

Rearranging this equation,

Mass of ethanol $\qquad$ = Mass fraction of ethanol x Total mass of solution

$\qquad\qquad\qquad\qquad$ = 35/100 x 500 = **175 g**

4.4. PERCENT MASS/VOLUME (M/V %)

This unit of concentration is generally only used with one solid and one liquid.

Note: m/v % = w/v %

In examples and exercises both m/v % and w/v % concentration units will be used.

Percent composition (m/v %) = Mass of Solute /Total volume of Solution x 100

For example, a 3 m/v % NaCl solution contains 3 g of NaCl for every 100 cm³ of solution. Given the different units in the numerator and denominator this is not a true percent. However, it is a good approximation given that the density of dilute aqueous solutions is generally close to 1 g/cm³, meaning that the volume of a solution in cm^3 is very nearly numerically equal to the mass of the solution in grams.

4.5. PERCENT BY VOLUME (V/V %)

Volume percent or volume/volume percent (v/v %) is used when preparing solutions of liquids. It is very easy to prepare a chemical solution using volume percent, but if you misunderstand the definition of this unit of concentration, you'll experience problems.

Volume percent (v/v %) is defined as:

$$v/v \% = [(\text{volume of solute})/(\text{volume of solution})] \times 100\%$$

Note that volume percent is relative to the volume of solution, not to the volume of the solvent. For example, lagers typically contain 4 v/v % ethanol. This means there are 4 cm^3 ethanol for every 100 cm^3 of lager.

Example 4.12: *A 400 cm^3 bottle of solution used in developing photographs contains 140 cm^3 of pure acetic acid. What is the v/v concentration of the solution ?*

Answer:

Concentration = Volume of Solute / Volume of Solution x 100
= 140/400 x 100 = **35 v/v %**

4.6. PARTS PER MILLION (PPM)

For very dilute solutions concentrations are often expressed in parts per million (ppm). 1 ppm is one part by weight, or volume, of solute in 1 million parts by weight or volume, of solution.

In weight/volume (m/v) terms, 1 ppm = 1 mg/L = 1mg/dm^3

In weight/weight (m/m) terms, 1 ppm = 1 mg/kg = 1 mg/kg

Example 4.13: *A solution has a concentration of 1.25 g/L. What is the concentration in ppm ?*

Answer:

Convert the mass in grams to milligrams (mg): 1.25 g = 1.25 x 1000 mg = 1250 mg

Rewrite the concentration in mg/L: 1250 mg/L = **1250 ppm**

Example 4.14: *A solution has a concentration of 0.5 mg/cm³. What is this concentration in ppm ?*

Answer:

ppm = mass of solute (mg) / volume solution (L)

Convert volume to litres: 1 cm³ = 1/1000 = 0.001 L

Concentration = mass of solute/volume = 0.5/0.001 = 500 mg/L = **500 ppm**

Example 4.15: *150 cm³ of an aqueous sodium chloride solution contains 0.0045g NaCl. Calculate the concentration of NaCl in ppm.*

Answer:

ppm = mass of solute (mg) / volume solution (L)

Convert the mass in grams to milligrams: mass NaCl = 0.0045 g = 0.0045 x 1000 = 4.5 mg

Convert volume to litres: 150 cm³ = 15/1000 = 0.15L

Concentration = mass of solute/volume = 4.5/0.15 = **30 ppm**

Example 4.16: *The body tissue of workers recycling electronic waste in India was found on average to contain 4 ppm of polychlorinated biphenyls (PCBs). What mass of PCBs is present in a 64 kg person ?*

Answer:

1ppm = 1 mg/kg

4 ppm = 4 mg/kg

Thus 1 kg body tissue contains 4 mg PCBs

Therefore, 64 kg will contain 64 x 4 mg PCBs = 256 **mg**

Exercise 4.4

a. *8.0 g of copper was used to make 100 g of an alloy. What is the concentration (% w/w) of copper in the alloy ?*

b. *An item of brass costume jewellery with a mass of 35.0 g was found to contain 1.7 g of zinc. Calculate the mass by percent of zinc in the brass ?*

c. *You have 200 g of a solution that contains 30 g of hydrochloric acid (HCl). What percentage of your solution is made up of hydrochloric acid (m/m%) ?*

d. *If I make a solution by adding water to 65 cm³ of ethanol until the total volume of the solution is 350 cm³, what's the percent by volume of ethanol in the solution ?*

e. *Solder flux typically contains 8 g of zinc chloride per 25 cm³ of aqueous hydrochloric acid. Determine the percent mass by volume of the zinc chloride in the flux*

f. *150 g of sodium chloride is present in 4 litres of a solution. Calculate the percent mass by volume of sodium chloride in the solution.*

g. *Suppose you have 70 g of sodium chloride salt (NaCl) in 250 cm³ of water. Express this as a (m/v %) solution.*

h. *A solution is prepared by dissolving 50.0 g of caesium chloride (CsCl) in 50.0 g water. Calculate the mass % of caesium chloride in the solution.*

i. *A solution is prepared by dissolving 125 g sucrose (C₁₂H₂₂O₁₁) in 135 g H₂O. Calculate the mass % of sucrose in the solution.*

j. *A 300g sample of river water contains 38 mg of lead. Express this concentration in ppm*

k. *0.425 g of lead sulfate was dissolved in 100 g of water. Calculate the concentration of lead present in ppm.*

l. *A 900.0 g sample of seawater was found to contain 6.7 x 10⁻³ g of Zn. Express this concentration in ppm.*

m. *Surgical spirit contains 70 % vv isopropyl alcohol (C₃H₇OH). What volume of pure C₃H₇OH is present in a 250 cm³ bottle of surgical spirit ?*

n. *The maximum concentration of fluoride ions permitted in UK drinking water is 1.5 ppm. Calculate the maximum mass of fluoride ions contained in a glass (250 cm³) of UK drinking water.*

o. *Birds of prey in the UK typically contain polyhalogenated aromatic* hydrocarbons (PHAHs) at an average concentration of 18.9 mg/kg. Calculate the mass PHAHs present in a chick weighing 0.6 kg

p. The urine sample for an athlete who tested positive for a banned substance was found contain thousand times the permitted the maximum acceptable level of 2 mg/L. What was the test result concentration in parts per million ?

q. Milk is categorized according to its fat content. Semi-skimmed milk contains 1.7 g of fat per 100 cm³. How much fat is present in a small carton (250cm³) of semi-skimmed milk ?

r. Water from a farm well was found to contain 55 ppm nitrate. Determine the mass of nitrate ions present in a 200 cm³ sample of the water.

s. Antifreeze is a 40 % v/v solution of ethylene glycol in water. Calculate the volume of ethylene glycol required to produce 5.8 L of antifreeze.

t. The water in a swimming pool was analysed for its chlorine content. It was found that 20 cm³ of a water sample contains 0.45 mg of free chlorine. What is the concentration of chlorine in ppm ?

ANSWERS

Exercise 4.1

Calculate the concentration of the following solutions containing:

Concentration = Number of Moles /Volume
Since Concentration is M,$i.e$mol/L (M), solution volumes must be in litres (L) or dm³

a. *2.65 g of dissolved sodium chloride (NaCl) in 200 cm³ of water.*

   Answer

   M_r [NaCl] = 58.5
   Number of moles in 2.65 g of NaCl = 2.65/58.5 = 0.453
   Concentration of 2.65 g NaCl in 200 cm³ (0.2 L) = number of moles/volume of solution = 0.453/0.2 = **0.226** M

b. *10.6 g of dissolved sodium carbonate (Na$_2$CO$_3$) in 500 cm³ of water.*

Answer

M_r [Na_2CO_3] = 106

Number of moles in 10.6 g of Na_2CO_3 = 10.6/106 = 0.1

Concentration of 10.6 g of Na_2CO_3 in 500 cm³ (0.5 L) = number of moles/volume of solution = 0.1/0.5 = **0.2 M**

c. *0.005 g of dissolved oxygen (O_2) in 100 cm³ of water.*

Answer

M_r [O_2] = 32

Number of moles in 0.005 g of O_2 = 0.005/32 = 1.56 x 10^{-4}

Concentration of 0.005 g of O_2 in 100 cm³ (0.1 L) = number of moles/volume of solution = 1.56 x 10^{-4} /0.1 = **1.56 x 10^{-3} M**

d. *6 g of dissolved calcium hydroxide {$Ca(OH)_2$} in 800 cm³ of water.*

Answer

M_r [$Ca(OH)_2$] = 74

Number of moles in 6 g of $Ca(OH)_2$ = 6/74 = 0.08

Concentration of 6 g $Ca(OH)_2$ in 800 cm³ (0.8 L) = number of moles/volume of solution = 0.08/0.8 = **0.1 M**

e. *0.34 g of dissolved sodium hypochlorite ($NaClO$) in 50 cm³ of water.*

Answer

M_r [$NaClO$] = 74.5

Number of moles in 0.34 g of $NaClO$ = 0.34/74 = 4.56 x 10^{-3}

Concentration of 0.34 g of $NaClO$ in 50 cm³ (50/1000 = 0.05 L) of water= number of moles/volume of solution = 4.56 x 10^{-3} /0.05 = **0.0912 M**

f. *1.2 g of dissolved sodium thiosulfate ($Na_2S_2O_3$) in 150 cm^3 of water.*

Answer

M_r $[Na_2S_2O_3]$ =158
Number of moles in 1.2 g of $Na_2S_2O_3$ = 1.2/158 = 7.6 x 10^{-3}
Concentration of 1.2 g $Na_2S_2O_3$ in 150 cm^3 (150/1000 = 0.15 L) = number of moles/volume of solution = 7.6 x 10^{-3}/0.15 = **0.051 M**

g. *2.356 g of dissolved in ammonium hydrogen carbonate {(NH_4)HCO_3} in 350 cm^3 of water.*

Answer

M_r $[(NH_4)HCO_3]$ = 79
Number of moles in 2.356 g of (NH_4)HCO_3 = 2.356/79 = 0.0298
Concentration of 2.356g of (NH_4)HCO_3 in 350 cm^3 (350/1000 = 0.35 L) = number of moles/volume of solution = 0.0298/0.35 = **0.085 M**

h. *0.3 g of dissolved calcium oxide (CaO) in 250 cm^3 of water.*

Answer

M_r [CaO] = 56
Number of moles in 0.3 g of CaO = 0.3/56 = 5.3 x 10^{-3}
Concentration of 0.3 g of CaO in 250 cm^3 (0.25 L) = number of moles/volume of solution = 5.3 x 10-3/0.25 =**0.021 M**

i. *0.45 g of dissolved oxalic acid ($C_2H_4O_2$) in 200 cm^3 of water.*

Answer

M_r $[C_2H_4O_2]$ = 60
Number of moles in 0.45 g of $C_2H_4O_2$ = 0.45/60 = 7.5 x 10^{-3}

Concentration of 0.45 g of $C_2H_4O_2$ in 200 cm³ (200/1000 = 0.2 L) = number of moles/volume of solution = 7.5 x 10^{-3}/0.2 =**0.0375 M**

j. *1.24 g of dissolved sodium hydrogen carbonate (NaHCO₃) in 75 cm³ of water.*

Answer

M_r [NaHCO$_3$] = 84
Number of moles in 1.24 g of NaHCO$_3$ = 1.24/84 = 0.0148
Concentration of 1.24g of NaHCO$_3$ in 75 cm³ (75/1000 = 0.075 L) = number of moles/volume of solution = 0.0148/0.075 =**0.197 M**

Exercise 4.2

What mass of solute must be dissolved in water to make:

a. *100 cm³ of 1 M sodium hydroxide (NaOH) ?*

Answer

M_r [NaOH] = 40
Number of moles in 100 cm³ (0.1 L) of 1.0 M NaCl = concentration x volume of solution = 1 x 0.1 = 0.1 moles
Mass of 0.1 moles of NaCl = number of moles x M_r = 0.1 x 40 = **4 g**

b. *500 cm³ of 2 M nitric acid (HNO₃) ?*

Answer

M_r = [HNO$_3$] = 63
Number of moles in 500 cm³ (0.5 L) of 2.0 M HNO$_3$ = concentration x volume of solution = 2 x 0.5 = 1 mole
Mass of 1 mole of HNO$_3$ = number of moles x M_r = **63 g**

c. *1 cm³ of 0.05 M silver nitrate (AgNO₃) ?*

Answer

M_r [$AgNO_3$] = 170
Number of moles in 1 cm^3 (0.001 L) of 0.05 M $AgNO_3$ = concentration x
volume of solution = 0.05 x 0.001 = 5 x 10^{-5} mole
Mass of 5 x 10^{-5} mole of $AgNO_3$ = number of moles x M_r = 5 x 10^{-5} x 170 =
0.0085 g

d. *25 cm^3 of 1.5 x 10^3 M sulfuric acid (H_2SO_4) ?*

Answer

M_r [H_2SO_4]= 98
Number of moles in 25 cm^3 (0.025 L) of 1.5 x 10^{-3} M H_2SO_4 = concentration x
volume of solution = 0.025 x 1.5 x 10^{-3} = 3.75 x 10^{-5}
Mass of 3.75 10^{-5} moles = number of moles x M_r = 3.75 10^{-5} x 98 = **3.67 x 10^{-3} g**

e. *17.4 cm^3 of 0.125 M oxalic acid ($C_2H_4O_2$) ?*

Answer

M_r [$C_2H_4O_2$] = 60
Number of moles in 17.4 cm^3 (0.0174 L) of 0.125 M = volume x concentration
= 0.0174 x 0.125 = 0.0022
Mass of moles of $C_2H_4O_2$ = number of moles x M_r = 0.0022 x 60 = **0.132 g**

f. *150 cm^3 of 0.12 M sucrose ($C_{12}H_{22}O_{11}$) ?*

Answer

M_r [$C_{12}H_{22}O_{11}$] = 342
Number of moles in 150 cm^3 (0.150 L) of 0.12 M = volume x concentration =
0.12 x 0.15 = 0.018 moles
Mass of 0.018 moles of $C_{12}H_{22}O_{11}$ = number of moles x M_r = 0.018 x 342 =

6.16 g

g. *23.5 cm³ of 0.00125 M barium sulfate (BaSO₄) ?*

Answer

M_r [$BaSO_4$] = 233
Number of moles in 23.5 cm³ (0.0235 L) of 0.00125 M = volume x concentration = 0.0235 x 0.00125 = 2.935 x 10^{-5}
Mass of 2.935 x 10^{-5} moles of $BaSO_4$ = number of moles x M_r = 2.935 x 10^{-5} x 233 = **6.84 x 10^{-3} g**

h. *12.4 cm³ of 0.101 M hydrochloric acid (HCl) ?*

Answer

M_r [HCl] = 36.5
Number of moles in 12.4 cm³ (0.0124 dm⁻³) of 0.101 M = volume x concentration = 0.0124 x 0.101 = 0.001252
Mass of 0.001252 moles of HCl = number of moles x M_r = 0.001252 x 36.5 = **0.0457 g**

i. *5 cm³ of 2 x 10⁻⁴ M ascorbic acid (C₆H₈O₆) ?*

Answer

M_r [$C_5H_8O_6$] = 176
Number of moles in 5 cm³ (0.005 L) of 2 x 10^{-4} M = volume x concentration = 0.005 x 2 x 10^{-4} = 1 x 10^{-6}
Mass of 1 x 10^{-6} moles of $C_5H_8O_6$ = number of moles x M_r = 1 x 10^{-6} x 176 = **1.76 x 10^{-4} g**

j. *12.3 cm³ of 1.03 x 10⁻⁴ M iron (III) chloride (FeCl₃) ?*

Answer

$M_r[FeCl_3] = 162$
Number of moles in 12.3 cm^3 (0.0123 L) of 1.03 x 10^{-4} M $FeCl_3$ = volume x concentration = 0.0123 x 1.03 x 10^{-4} = 1.27 x 10^{-6}
Mass of 1.27 x 10^{-6} moles of $FeCl_3$ = number of moles x M_r = 162 x 1.27 x 10^{-6}
= **2.05 x 10^{-4} g**

Exercise 4.3

How many moles are present in the following solutions

$$\text{number of moles} = \text{volume (L)} \times \text{concentration (M)}$$
$$\text{volume (L)} = \text{volume (cm}^3)/ 1000$$

a. *25 cm^3 of 1 M NaOH ?*

Answer

Number of moles of NaOH = 1 x 25/1000 = **0.025**

b. *7.5 litres of 5 M H_2SO_4 ?*

Answer

Number of moles of H_2SO_4 = 5 x 7.5 = **37.5**

c. *50 cm^3 of 0.101 M NaCl ?*

Answer

Number of moles of NaCl = 0.101 x 50/1000 = **5.05 x 10^{-3}**

d. *12.7 cm^3 of 0.035 M $K_2Cr_2O_7$*

Answer

Number of moles of $K_2Cr_2O_7$ = 0.035 x 12.7/1000 = **4.445 x 10^{-4}**

e. *2.6 cm^3 of 0.102 M H_2O_2*

Answer

Number of moles of H_2O_2 = 0.102 x 2.6/1000 = **2.65 x 10^{-4}**

f. *32.8 cm^3 of 0.033 M HNO_3*

Answer

Number of moles of HNO_3 = 0.033 x 32.8/1000 = **1.08 x 10^{-3}**

g. *12.8 cm^3 of 0.65 M $NaMnO_4$*

Answer

Number of moles of $NaMnO_4$ =0.65 x 12.8/1000 = **8.32 x 10^{-3}**

h. *42.13 cm^3 of 0.05 M $AgNO_3$*

Answer

Number of moles of $AgNO_3$ = 0.05 x 42.13/1000= **2.1 x 10^{-3}**

i. *5.6 litres of 1.35 M NaCl*

Answer

Number of moles of NaCl = 1.35 x 5.6 = **7.56**

j. *23.9 cm³ of 1.56 x 10⁻³ MKI*

Answer

Number of moles of KI =1.56 x 10⁻³ x 23.9/1000 = **3.73 x 10⁻⁵**

Exercise 4.4

a. *8.0 g of copper was used to make 100 g of an alloy. What is the concentration (% w/w) of copper in the alloy ?*

Answer

w/w %   = mass of solute/total mass x 100
         = 8/100 x 8
         = **8 w/w %**

b. *An item of brass costume jewellery with a mass of 35.0 g was found to contain 1.7 g of zinc. Calculate the mass by percent of zinc in the brass ?*

Answer

m/m %   = mass of solute/ total mass x 100
         = 1.7 /35 x 100
         = **4.86 m/m %**

c. *You have 200 g of a solution that contains 30 g of hydrochloric acid (HCl), what percentage of your solution is made up of hydrochloric acid (m/m %) ?*

Answer

m/m %   = mass of solute/ total mass
         = 30/200 x 100
         = **15 m/m %**

d. *If I make a solution by adding water to 65 cm³ of ethanol until the total volume of the solution is 350 cm³, what's the percent by volume of ethanol in the solution ?*

Answer

v/v %   = volume of solute/ volume of solution x 100
            = 65/350 x 100
            = **18.6 v/v %**

e. *Solder flux typically contains 8g of zinc chloride per 25 cm³ of aqueous hydrochloric acid. Determine the percent mass by volume (m/v %) of the zinc chloride in the flux.*

Answer

m/v %   = mass of solute / volume of solution x 100
            = 8 / 25 x 100
            = **32 m/v %**

f. *150 g of sodium chloride is present in 4 litres of a solution. Calculate the percent mass by volume (m/v %) of sodium chloride in the solution.*

Answer

m/v %   = mass of solute / volume of solution x 100
            = 150g / 4000 x 100
            = **8.5 m/v %**
Note: If mass is in g then volume must be in cm³

g. *Suppose you have 70 g of sodium chloride salt (NaCl) in 250mL of water. Express this as a (m/v)% solution.*

Answer

m/v %   = mass of solute / volume of solution x 100
               = 70/250 x 100
               = **28 m/v %**

h. *A solution is prepared by dissolving 50.0 g of caesium chloride (CsCl) in 50.0 g water. Calculate the mass % of caesium chloride in the solution.*

Answer

Mass %   = Mass of CsCl/ Total Mass of Solution x 100
               = 50/100 x 100
               = **50 %**

i. *A solution is prepared by dissolving 125 g sucrose ($C_{12}H_{22}O_{11}$) in 135 g H_2O. Calculate the mass % of sucrose in the solution.*

Answer

Mass %   = Mass of Sucrose/Total Mass of Solution x 100
               = 125/(125+135) x 100
               = **48 %**

j. *A 300 g sample of river water contains 38 mg of lead. Express this concentration in ppm.*

Answer

ppm = mg/kg
If 300 g contains 38 mg Pb, then 1 kg (*i.e* 1000g) contains 1000/300 x 38 = 126.7 mg
Thus 126.7 mg in 1 kg = **126.7 ppm**

k. *0.425 g of lead sulfate was dissolved in 100 g of water. Calculate the concentration of lead present in ppm.*

Answer

ppm = mg/kg
Convert g to mg: 0.425 g = 0.425 x 1000 = 425 mg
Concentration (ppm) = mass of solute (mg) / mass of solution (kg) = 425/0.1 = **4250 ppm**

l. A 900.0 g sample of sea water is found to contain 6.7 x 10^{-3} g Zn. Express this concentration in ppm.

Answer

ppm = mg/kg
Convert mass of solute to mg: 6.7 x 10^{-3} x 1000 = 6.7 mg
Mass of solution = 0.90 kg
Concentration = mass of solute (mg)/volume of solution (kg) = 6.7/0.9 = **7.4 ppm**

m. *Surgical spirit contains 70 %vv isopropyl alcohol (C_3H_7OH). What volume of pure C_3H_7OH is present in a 250 cm^3 bottle of surgical spirit ?*

Answer

Concentration (v/v)	= Volume of C_3H_7OH/Total Volume of Solution x 100
Rearranging, Volume of C_3H_7OH	= {Concentration (% v/v) x Total Volume of Solution} /100
	= (70 x 250)/100
	= **17.5 cm^3**

n. *The maximum concentration of fluoride ions permitted in UK drinking water is 1.5 ppm. Calculate the maximum mass of fluoride ions contained in a glass (250 cm³) of UK drinking water.*

Answer

1ppm = 1 mg /L
Thus 1.5 ppm   = 1.5 mg fluoride in 1 L (1000 cm³)
Mass of fluoride   = concentration (ppm) x volume (L)
                              = 1.5 x 0.25 = **0.375 mg**

o. *Birds of prey in the UK typically contain polyhalogenated aromatic hydrocarbons (PHAHs) at an average concentration of 18.9 mg/kg. Calculate the mass PHAHs present in a chick weighing 0.6 kg*

Answer

1 kg contains 18.9 mg of PCBs
Therefore, 0.6 kg must contain 18.9 x 0.6 mg = **11.34 mg**

p. *The urine sample for an athlete who tested positive for a banned substance was found contain thousand times the permitted the maximum acceptable level of 2 mg/L. What was the test result concentration in parts per million ?*

Answer

Maximum acceptable level = 2 mg/L = 2 ppm
1000 times maximum acceptable level = 2 x 1000 = **2,000 ppm**

q. *Milk is categorized according to its fat content. Semi-skimmed milk contains 1.7 g of fat per 100 cm³. How much fat is present in a small carton (250cm³) of semi-skimmed milk ?*

Answer

$100 \ cm^3$ of milk contains 1.7 g of Milk Fat

Therefore, $250 \ cm^3$ contains 250/100 x 1.7 = **4.2 g Milk Fat**

r. *Water from a farm well was found to contain 55 ppm nitrate. Determine the mass of nitrate ions present in a 200 cm^3 sample of the water.*

Answer

1 ppm = 1 mg/L

Convert volume to litres, $200 \ cm^3$ = 200/100 = 0.2 L

55 ppm = 55 mg/L

Therefore, 0.2 L will contain 0.2 x 55 ppm = **11 ppm**

s. *Antifreeze is a 40% v/v solution of ethylene glycol in water. Calculate the volume of ethylene glycol required to produce 5.8 L of antifreeze.*

Answer

Concentration (% v/v) = Volume of Solute/Total Volume of Solution x 100

Rearranging, Volume of Solute = {Concentration (% v/v) x Total Volume of Solution}/100

Therefore, Volume of ethylene glycol = 40 x 5.68 = **2.272 L**

t. *The water in a swimming pool was analysed for its chlorine content. It was found that 20 cm^3 of a water sample contains 0.45 mg of free chlorine. What is the concentration of chlorine in ppm ?*

Answer

1 ppm = 1 mg/L

Concentration (ppm) = Mass (mg)/volume (L)

Chlorine concentration = 0.45/0.020 = **22.5 ppm**

CHAPTER 5

Volumetric Analysis

Keywords: Acid-Base titrations, Back titrations, Calculations, Concentration, End point, Redox titration, Standard solutions, Volumetric analysis.

5.1 INTRODUCTION

A solution of accurately known concentration is called a **standard solution**. Standard solutions are used in volumetric (titrimetric) analysis, to determine the concentration an unknown solution. The volume of one solution that will react with a known volume of a standard solution (titrant) is determined. The point at which the exact amount of titrant added to just react with all of the other reagent present is called the **end point** or **equivalence point**. Indicators, normally added to the solution of known volume, which change colour, are often used to determine the end point.

Volumetric analysis is widely used to determine the concentration of a broad range variety of parameters including alkalinity, acidity, total hardness and chloride levels.

Titrations can be categorised based on chemical reactions:

> **Acid–Base** titrations involve the exact neutralisation of an acid or base with an acid or base of known concentration, thus allowing the concentration of the unknown acid or base solution to be determined.
> **Redox** titrations can be used to determine oxidizing or reducing agents in a solution. The reducing or oxidizing agent is used as the titrant against the other agent.
> **Back Titrations** are used where analytes are either partially soluble or too slow to give a reaction. A known amount of excess reagent is used. The remaining excess reagent is then titrated with another second reagent to determine how much of the excess reagent was used in the first

Nigel P. Freestone
All rights reserved-© 2016 Bentham Science Publishers

titration, allowing the original analyte's concentration to be determined.

Example 5.1: *Chloride concentrations in water can be determined by titration against standardised silver nitrate solutions. A 50 cm³ water sample was titrated against 0.05 M silver nitrate solution, using potassium chromate as the indicator. This indicator changes colour when all the chloride has been precipitated out of solution as silver chloride. This colour change occurred after the addition of 10.9 cm³ of 0.05 M silver nitrate. What is the concentration of chloride ions in the water sample?*

Step 1: Write the balanced chemical equation, insert the information given in the question and identify what you are trying to calculate.

	$Cl^-_{(aq)}$	+	$AgNO_{3(aq)}$	$\rightarrow$	$AgCl_{(s)} + NO_{3\ (aq)}^-$
Reaction coefficients	1		1		
Volume (cm³)	50 (0.05 L)		10.9		
No. of Moles					
Concentration (M)	?		0.05		

Please note that volumetric analysis (titrations) are predominantly concerned with the reaction coefficient (stoichiometric) relationship between the reactants. The products are included in the above table for the sake of completeness.

Step 2: If two pieces of information from volume, number of moles, and concentration are known for a given species, the third can be calculated using:

Concentration = Number of Moles / Volume of solution
c = n / v
OR
Number of Moles = Volume of Solution x Concentration
n = c x v
OR
Volume of Solution = Number of Moles / Concentration
v = n / c

Units: Concentration: M,

       Volume: Litres (L) or {volume $(cm^3)/1000$}

Number of moles in 10.9 cm^3 0.05 M $AgNO_3$ = concentration x volume

$$= 0.05 \times 10.9/1000 = \mathbf{5.45 \times 10^{-4}}$$

	$Cl^-_{(aq)}$	+	$AgNO_{3(aq)}$	→	$AgCl_{(s)} + NO_{3\ (aq)}^-$
Reaction Coefficients	1		1		
Volume (cm^3)	50 (0.05 L)		10.9		
No. of Moles			5.45×10^{-4}		
Concentration (M)	?		0.05		

Step 3: At the end point the amount of each species present is in accordance with the reaction coefficient (stoichiometric) relationship of the balanced chemical equation for the reaction. In this example, the number of moles of Cl^- in 50 cm^3 of tap water = the number of moles Ag^+ in 10.89 cm^3 of the standard solution = 5.45×10^{-4}

	$Cl^-_{(aq)}$	+	$AgNO_{3(aq)}$	→	$AgCl_{(s)} + NO_{3\ (aq)}^-$
Reaction Coefficients	1		1		
Volume (cm^3)	50 (0.05 L)		10.9		
No. of Moles	5.45×10^{-4}		5.45×10^{-4}		
Concentration (M)	?		0.05		

Step 4: Since we know both the number of moles of Cl^- and its volume, we can calculate its concentration.

     Concentration of Cl^- = number of moles / volume (L)

                        $= 5.45 \times 10^{-4}/0.05$

                        $= 0.0109$ M

	$Cl^-_{(aq)}$	+	$AgNO_{3(aq)}$	→	$AgCl_{(s)} + NO_{3\ (aq)}^-$
Reaction coefficients	1		1		
Volume (cm^3)	50 (0.05 L)		10.9		
No. of Moles	5.45×10^{-4}		5.45×10^{-4}		
Concentration (M)	0.0109		0.05		

Answer:

Concentration of Cl⁻ in tap water = **0.0109 M**

Example 5.2: *A solution made from anhydrous sodium carbonate contained 2.6061 g in exactly 250 cm³ of water. Using methyl orange indicator, titration of 25.0 cm³ of this solution required 18.7 cm³ of a hydrochloric acid solution for complete neutralisation. What is the concentration of the acid?*

Step 1: Fill in the information given in the question

	$Na_2CO_{3(aq)}$	+	$2HCl_{(aq)}$	→	$2NaCl + CO_2 + H_2O$
Reaction Coefficients	1		2		
Volume (cm³)	25		18.7		
No. of Moles					
Concentration (M)			?		

Step 2: You need to know two pieces of information (volume, number of moles or concentration) about one of the species and one piece of information about another to solve volumetric (titrimetric) problems. If this information is not given directly in the question then it can be calculated from the information provided - in this example the concentration of Na_2CO_3.

M_r $Na_2CO_3 = 106$
Number of moles in 2.6061 g of Na_2CO_3 = mass/M_r = 2.6061/106 = 0.0246
Therefore, the number of moles of Na_2CO_3 in 250 cm³ (0.25 L) = 0.0246
25 cm³ of the Na_2CO_3 solution contains 0.0246 x 25/250 = 2.46 x 10⁻³ moles of Na_2CO_3
Concentration (*i.e.* number if moles in 1000 cm³) = number of moles /volume = 0.0246/0.25 = 0.098 M

	$Na_2CO_{3(aq)}$	+	$2HCl_{(aq)}$	→	$2NaCl + CO_2 + H_2O$
Reaction Coefficients	1		2		
Volume (cm³)	25 (0.025 L)		18.7 (0.0187 L)		

Table contd.....

	$Na_2CO_{3(aq)}$	+	$2HCl_{(aq)}$	$\rightarrow$	$2NaCl + CO_2 + H_2O$
No. of Moles	2.46×10^{-3}				
Concentration (M)	0.098		?		

Step 3: At the end point (*i.e.* when the indicator changes colour) there are 2 moles of HCl for every mole of Na_2CO_3.

Number of moles of HCl at the end point = 2 x the number of moles of
$$Na_2CO_3$$
$$= 2 \times 2.46 \times 10^{-3}$$
$$= 4.92 \times 10^{-3} \text{ moles}$$

	$Na_2CO_{3(aq)}$	+	$2HCl_{(aq)}$	$\rightarrow$	$2NaCl + CO_2 + H_2O$
Reaction Coefficients	1		2		
Volume (cm³)	25 (0.025 L)		18.7 (0.0187 L)		
No. of Moles	0.0246		4.92×10^{-3}		
Concentration (M)	0.098		?		

Step 4:

Thus the number of moles of HCl in 18.7 cm³ = 4.92×10^{-3} moles
Concentration is defined as the number of moles in 1 L *i.e.* = number of moles / volume
Convert volume from cm³ to L, 18.7 cm³ = 18.7/1000 = 0.0187 L
Therefore concentration of HCl = number of moles/volume = 4.92×10^{-3} /0.0187 = 0.263 M

	$Na_2CO_{3(aq)}$	+	$2HCl_{(aq)}$	$\rightarrow$	$2NaCl + CO_2 + H_2O$
Reaction Coefficients	1		2		
Volume (cm³)	25 (0.025 L)		18.7 (0.0187 L)		
No. of Moles	0.0246		4.92×10^{-3}		
Concentration (M)	0.098		0.263		

Answer:

HCl concentration = **0.263 M**

Example 5.3: *The purity of magnesium oxide, which is not very soluble in water, can be determined by use of the 'back titration' method. A solution was made by completely dissolving 4.06 g of impure magnesium oxide in 100 cm³ of 2.0M hydrochloric acid. 19.7 cm³ of 0.2 M sodium hydroxide was required to neutralize the excess acid. The second titration is called a 'back-titration', is used to determine the unreacted acid*

a) Write equations for the two neutralisation reactions.

Answer:

MgO + excess HCl → $MgCl_2$ + H_2O + unreacted HCl
$NaOH$ + HCl → NaCl + H_2O

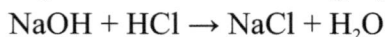

b) How many moles of hydrochloric acid were added to the magnesium oxide?

Answer:

Number of moles in 100 cm³ of 2M HCl = volume x concentration = 100/1000 x 2 = 0.2 moles

c) Determine the number of moles of excess hydrochloric acid titrated.

Answer:

Number of moles in 19.7 cm³ (0.019 L) of 0.2M HCl = volume x concentration = 0.0197 x 0.2 = 0.0197 moles

d) Calculate the moles of hydrochloric acid reacting with the magnesium oxide.

Answer:

Number of mole reacting with MgO = 0.2 - 0.01986 = 0.1804 moles

e) Calculate the moles and mass of magnesium oxide that reacted with the initial hydrochloric acid.

Answer:

M_r MgO = 40

According to the reaction coefficients, 2 moles of HCl react with 1 mole of MgO

Therefore, the number of moles of MgO = 0.1804 = 0.0902

Mass of MgO = number of moles x M_r = 0.0902 x 40 = 3.608 g

f) What is the percentage purity of the magnesium oxide?

Answer:

% purity = 3.608/4.06 x 100 = 88.9%

Example 5.4: *A 0.400g sample of impure sodium hydrogen carbonate was dissolved in 100.0 cm³ of water and titrated with 0.200 M hydrochloric acid. 23.75 cm³ of acid was required for complete neutralisation. Calculate the mass of sodium hydrogen carbonate titrated and hence the purity of the sample.*

Answer:

	$Na_2CO_{3(aq)}$	+	$2HCl_{(aq)}$	→	$2NaCl + CO_2 + H_2O$
Reaction coefficients	1		1		
Volume (cm³)	100 (0.1L)		23.75		
No. of Moles	4.75 x 10⁻³		4.75 x 10⁻³		
Concentration (M)	**0.0476**		0.2		

M_r NaHCO$_3$ = 84

Number of moles in 23.75 cm³ of 0.2M HCl	= volume x concentration = 23.75/1000 x 0.2 = 4.75 x 10⁻³
Volume	= 100 cm³ = 0.1 L
Concentration of prepared NaHCO₃ solution	= number of moles / volume = 4.76 x 10⁻³ /0.1 = 0.0476 M
At the end point the number of moles of NaHCO₃	= number of moles of HCl = 4.75 x 10⁻³
Mass of 4.75 x 10⁻³ of NaHCO₃	= number of moles x M_r = 4.75 x 10⁻³ x 84 = **0.399 g**
Purity (%)	= 0.399/0.400 x 100 = **99.75%**

Example 5.5: *The redox reaction between permanganate ions and iron(II) ions is:*

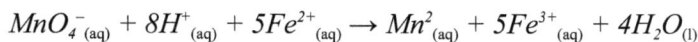

$$MnO_4^-{}_{(aq)} + 8H^+{}_{(aq)} + 5Fe^{2+}{}_{(aq)} \rightarrow Mn^2{}_{(aq)} + 5Fe^{3+}{}_{(aq)} + 4H_2O_{(l)}$$

What volume (in cm^3) of permanganate solution of concentration 0.04M would react exactly with 30 cm^3 of a solution of iron(II) which has a concentration of 0.3 M?

Answer:

	MnO_4^-	+	$8H^+$	+	$5Fe^{2+}$	→	$Mn^{2+} + 5Fe^{3+} + 4H_2O$
Reaction Coefficients	1		8		5		
Volume (cm^3)	4.5 (4.5 x 10^{-3} L)				30 (0.03 L)		
Moles	1.8 x 10^{-3}				9 x 10^{-3}		
Concentration (M)	0.4				0.3		

Number of moles in $30cm^3$ of 0.3M Fe^{2+} = volume x concentration
$$= 30/1000 \times 0.3 = 9 \times 10^{-3}$$

According to the reaction coefficients, at the end point number of moles of MnO_4^- = 1/5 x number of moles of Fe^{2+}

$$= 1/5 \times 9 \times 10^{-3}$$
$$= 1.8 \times 10^{-3}$$

Volume of $KMnO_4$ = number of moles x concentration
$$= 1.8 \times 10^{-3} /0.4 \times 1000 = \textbf{4.5 cm}^3 \textbf{ (4.5 x } 10^{-3} \textbf{ L)}$$

Example 5.6: *Succinic acid reacts with dilute sodium hydroxide according to the following equation:*

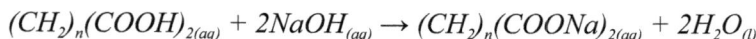

$$(CH_2)_n(COOH)_{2(aq)} + 2NaOH_{(aq)} \rightarrow (CH_2)_n(COONa)_{2(aq)} + 2H_2O_{(l)}$$

A solution was made by completely dissolving 2.0 g of succinic acid in water and the solution made up to 250 cm^3. 18.4 cm^3 of this solution was required to neutralize 25 cm^3 of 0. 1M NaOH. Use this information to determine both the molecular formula of the acid and hence the value of n.

Answer:

	$(CH_2)_n(COOH)_{2(aq)}$	+	$2NaOH_{(aq)}$	→	$(CH_2)_n(COONa)_2 + 2H_2O$
Reaction coefficients	1		2		
Volume (cm³)	18.4 (0.0184 L)		25 (0.025 L)		
Moles	1.25×10^{-3}		2.5×10^{-3}		
Concentration (M)	0.68		0.1		

Number of moles of NaOH in 25 cm³ of 0.1M NaOH

$= \text{volume x concentration}$
$= 25/1000 \times 0.1$
$= 2.5 \times 10^{-3}$

At the end point, the number of moles of $(CH_2)_n(COOH)_2$ in 18.4 cm³

$= 0.5 \text{ x number of moles of NaOH}$
$= 0.5 \times 2.5 \times 10^{-3} = 1.25 \times 10^{-3}$

Number of moles of $(CH_2)_n(COOH)_2$ in 250 cm³

$= 250/18.7 \times 1.25 \times 10^{-3}$
$= 0.017 \text{ moles}$

Thus 0.017 moles of $(CH_2)_{n(}COOH)_2$ has a mass of 2 g

$M_r (CH_2)_n(COOH)_2 = \text{mass / number of moles} = 2 / 0.017 = \textbf{118}$

$M_r CH_2 = 14$

$M_r (COOH)_2 = 90$

Thus $14n + 90 = 118$

Rearranging, $14n = 118 - 90$:

$n = 28/14 = \textbf{2}$

Exercise 5.1

a) *If it takes 24 cm³ of 0.1 M NaOH to neutralize 20 cm³ of an HCl solution, what is the concentration of the HCl?*

b) *If it takes 1 cm³ of 0.05 M HNO₃ to neutralize 25 cm³ of NaOH solution, what is the concentration of the NaOH solution?*

c) *If it takes 50 cm³ of 0.5 M KOH solution to completely neutralize 125 cm³ of sulfuric acid solution (H_2SO_4), what is the concentration of the H_2SO_4 solution?*

d) *An ammonia solution was reacted with sulfuric acid as shown in the equation below.*

$$2NH_{3(aq)} + H_2SO_{4(aq)} \rightarrow (NH_4)_2SO_{4(aq)}$$

Calculate the concentration of the ammonia solution given that it takes 30.8 cm³ of a 1.24 M solution of sulfuric acid to neutralize 25 cm³ of this ammonia solution for complete reaction.

e) A solution was prepared by diluting a 50 cm³ sulfuric acid sample to 1 litre. 20cm³ of this solution was found on titration, to completely neutralize 25.0 cm³ of 1.00 M aqueous sodium hydroxide.

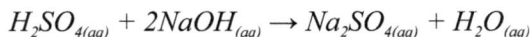

$$H_2SO_{4(aq)} + 2NaOH_{(aq)} \rightarrow Na_2SO_{4(aq)} + H_2O_{(aq)}$$

Calculate the concentration of the original concentrated sulfuric acid solution.

f) How many grams of Ca(OH)₂ are required to neutralize 52.68 cm³ of a 0.750 M H₂SO₄ solution ?

g) What volume of 1.0M sulfuric acid will be needed to neutralize 25.00 cm of 0.8M sodium hydroxide solution?

h) Calculate the volume of 0.10M sodium hydroxide solution needed to neutralize 20.00 cm³ of 0.25M hydrochloric acid.

i) 100 cm³ of a magnesium hydroxide solution required 4.5 cm³ of 0.10 M sulfuric acid for complete neutralisation. Calculate the concentration of the magnesium hydroxide in grams per litre.

j) Titration of 25.0 cm³ of a saturated calcium hydroxide solution required 35.45 cm³ of a 0.24 M hydrochloric acid solution for complete neutralisation. Calculate the molarity of the calcium hydroxide.

Exercise 5.2

a) If it takes 50 cm³ of 0.5 M KOH solution to completely neutralize 125 cm³ of sulfuric acid solution (H₂SO₄), what is the concentration of the H₂SO₄ solution?

b) Using phenolphthalein indicator, titration of 25.0 cm³ of 0.250M sodium hydroxide required 22.5 cm³ of a hydrochloric acid solution for complete neutralisation. What is the concentration of HCl?

c) A solution made from pure barium hydroxide contained 2.74 g in exactly 100 cm³ of water. Using phenolphthalein indicator, titration of 20.0 cm³ of this solution required 18.7 cm³ of a hydrochloric acid solution for complete neutralisation. What is the concentration of the acid? Aᵣ: Ba = 137, O = 16, H = 1.

d) A 100 cm³ sample of water is titrated against 0.0334 M Na₄EDTA solution, and requires 8.84 cm³ to reach the end point. Calculate the concentration of CaCO₃ in the water sample, given the following equation

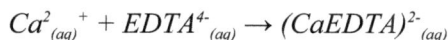

$$Ca^{2+}_{(aq)} + EDTA^{4-}_{(aq)} \rightarrow (CaEDTA)^{2-}_{(aq)}$$

e) What volume of 0.02 M KMnO₄ solution will react with 20cm³ of 0.1 M Fe²⁺(aq)? The reaction is:-

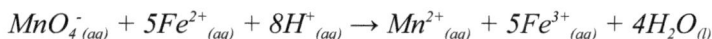

$$MnO_{4\ (aq)}^{-} + 5Fe^{2+}_{(aq)} + 8H^{+}_{(aq)} \rightarrow Mn^{2+}_{(aq)} + 5Fe^{3+}_{(aq)} + 4H_2O_{(l)}$$

f) 1.75 g of a hydrated sodium carbonate (Na₂CO₃.xH₂O) sample was dissolved in water and the volume made up to 250 cm³. 25.0 cm³ of this solution was found on titration, to require 24,5 cm³ of 0. 1M HCl according the following equation.

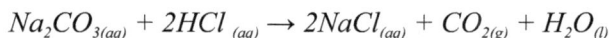

$$Na_2CO_{3(aq)} + 2HCl_{(aq)} \rightarrow 2NaCl_{(aq)} + CO_{2(g)} + H_2O_{(l)}$$

Calculate the value of x given the equation

g) 25 cm³ of a sample of vinegar (CH₃COOH) was dissolved in water, the resulting volume was made up to 250 cm³. 13.9 cm³ of this solution were required to neutralize 25 cm³ of 0. 1M NaOH. Calculate the molarity of the original vinegar solution and its concentration in g/L.

h) On titration, 0.2640 g of sodium oxalate dissolved in a flask required 30.74 cm³ of potassium permanganate to reach the end point. The equation for this reaction is:

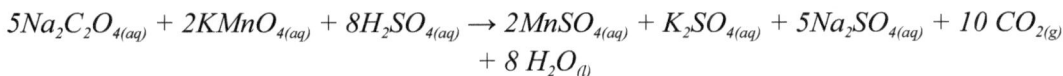

$$5Na_2C_2O_{4(aq)} + 2KMnO_{4(aq)} + 8H_2SO_{4(aq)} \rightarrow 2MnSO_{4(aq)} + K_2SO_{4(aq)} + 5Na_2SO_{4(aq)} + 10\ CO_{2(g)}$$
$$+\ 8\ H_2O_{(l)}$$

     i. *Determine the number of moles of oxalate in the flask.*
     ii. *How many moles of potassium permanganate were required to reach the end point?*
     iii. *Calculate the concentration of the potassium permanganate?*

i) A 2.5 g sample of ethanedioic acid, H₂C₂O₄.nH₂O, was dissolved in water and the solution made up to 250 cm³. 15.8 cm³ of this solution required to 25 cm³ of 0.1 M NaOH for complete neutralization. Calculate the value of n given that ethanedioic acid reacts with NaOH in a 1:2 ratio.

j) A 3.50 g sample of lawn sand containing an iron (II) salt was shaken with dilute H_2SO_4. This solution required 25.00 cm^3 of 0.0200 M potassium manganate (VII) for complete oxidation of the Fe (II) ions in the solution back to Fe(III) ions. Using the equation given below, determine the percentage of Fe(II) ions (by mass) in the lawn sand sample.

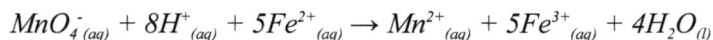

$$MnO_4^-{}_{(aq)} + 8H^+{}_{(aq)} + 5Fe^{2+}{}_{(aq)} \rightarrow Mn^{2+}{}_{(aq)} + 5Fe^{3+}{}_{(aq)} + 4H_2O_{(l)}$$

ANSWERS

Exercise 5.1

a) If it takes 24 cm^3 of 0.1 M NaOH to neutralize 20 cm^3 of an HCl solution, what is the concentration of the HCl?

Answer:

	$NaOH_{(aq)}$	+	$HCl_{(aq)}$	$\rightarrow$	$NaCl_{(aq)} + H_2O_{(l)}$
Reaction coefficients	1		1		
Volume (cm³)	24 (0.024 L)		20 (0.02 L)		
No. of Moles	2.4 x 10⁻³		2.4 x 10⁻³		
Concentration (M)	0.1		**0.12**		

Number of moles in 24 cm^3 of 0.1M NaOH = volume x concentration = 24/1000 x 0.1 =2.4 x 10⁻³

At the end point the number of moles of NaOH in 24 cm^3 = number of moles of HCl in 20cm^3

Therefore, number of moles of HCl in 20cm^3 (0.02 L) = 2.4 x 10⁻³

Concentration   = number of moles / volume
          = 2.4 x 10⁻³ / 0.02 = **0.12 M**

b) If it takes 1 cm^3 of 0.05 M HN0₃ to neutralize 25 cm^3 of NaOH solution, what is the concentration of the NaOH solution?

Answer:

	$NaOH_{(aq)}$	+	$HNO_{3(aq)}$	→	$2NaNO_{3(aq)} + H_2O_{(l)}$
Reaction coefficients	1		1		
Volume (cm³)	25 (0.025 L)		1 (0.001 L)		
No. of Moles	5×10^{-5}		5×10^{-5}		
Concentration (M)	2×10^{-3}		0.05		

Number of moles in 1 cm³ (0.001 L) of 0.05M HNO_3 = volume x concentration = 0.001 x 0.05 = 5×10^{-5}

At the end point the number of moles of HNO_3 in 1 cm³ = number of moles of NaOH in 25cm³

Therefore, number of moles in 25 cm³ (0.025 L) of NaOH = 5×10^{-5}

Concentration = number of moles / volume

$$= 5 \times 10^{-5} / 0.025$$
$$= \mathbf{2 \times 10^{-3} \ M}$$

c) If it takes 50 cm³ of 0.5 M KOH solution to completely neutralize 125 cm³ of sulfuric acid solution (H_2SO_4), what is the concentration of the H_2SO_4 solution?

Answer:

	$2KOH_{(aq)}$	+	$H_2SO_{4(aq)}$	→	$K_2SO_{4(aq)} + H_2O_{(l)}$
Reaction coefficients	2		1		
Volume (cm³)	50 (0.05 L)		125 (0.125 L)		
No. of Moles	0.025		0.0125		
Concentration (M)	0.5		0.1		

Number of moles in 50 cm³ 0.5 M of KOH = volume x concentration 50/1000 x 0.5 = 0.025

At the end point the number of moles of KOH in 50 cm³ = 0.5 x number of moles of H_2SO_4 in 125 cm³

Therefore, number of moles of KOH in 125 cm³ (0.125 L) = $^1/_2$ x 0.025 = 0.0125

Concentration of KOH = number of moles / volume = 0.0125 / 0.125 = **0.1 M**

d) An ammonia solution was reacted with sulfuric acid as shown in the equation below.

$$2NH_{3(aq)} + H_2SO_{4(aq)} \rightarrow (NH_4)_2SO_{4(aq)}$$

Calculate the concentration of the ammonia solution given that it takes 30.8 cm³ of a 1.24 M solution of sulfuric acid to neutralize 25 cm³ of this ammonia solution for complete reaction.

Answer:

	$2NH_{3(aq)}$	+	$H_2SO_{4(aq)}$	→	$(NH_4)_2SO_4$
Reaction coefficients	1		1		
Volume (cm³)	25 (0.025 L)		30.8 (0.038 L)		
No. of Moles	0.0764		0.0382		
Concentration (M)	**3.06**		1.24		

Number of moles in 30.8 cm³ of 1.24 M H_2SO_4 = volume x concentration
= 30.8/1000 x 1.24 = 0.0382
At the end point, the number of moles of NH_3 = 2 x number of mole of H_2SO_4 = 0.0382 x 2 = 0.0764
Thus number of moles of NH_3 in 25 cm³ (0.025 L) = 0.076
Concentration of NH^3 = number of moles / volume = 0.076 /0.025 = **3.06 M**

e) A solution was prepared by diluting a 50 cm³ sulfuric acid sample to 1 litre. 20cm³ of this solution was found on titration, to completely neutralize 25.0 cm³ of 1.00 M aqueous sodium hydroxide.

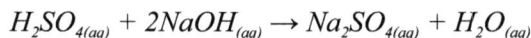

$$H_2SO_{4(aq)} + 2NaOH_{(aq)} \rightarrow Na_2SO_{4(aq)} + H_2O_{(aq)}$$

Calculate the concentration of the original concentrated sulfuric acid solution.

Answer:

	$H_2SO_4(aq)$	+	2NaOH(aq)	→	$Na_2SO_4 + H_2O$
Reaction coefficients	1		2		

Table contd.....

	H₂SO₄(aq)	+	2NaOH(aq)	→	Na₂SO₄ + H₂O
Volume (cm³)	20 (0.02L)		25 (0.025 L)		
No. of Moles	0.0125		0.025		
Concentration (M)	1		2		
Reaction coefficients	0.625		1		

Number of moles in 25 cm³ of 1 M NaOH = volume x concentration = 25/1000 x 1 = 0.025

At the end point, the number of moles of H_2SO_4 = 0.5 x number of mole of NaOH = 0.5 x 0.025 = 0.0125

Thus number of moles of H_2SO_4 in 20 cm³ (0.02 L) = 0.0125

Concentration of H_2SO_4 = number of moles / volume = 0.0125/0.02 = 0.625 M

Therefore, concentration of the original H_2SO_4 sample = 1000/50 x 0.625 = **12.5 M**

f) How many grams of Ca(OH)₂ are required to neutralize 52.68 cm³ of a 0.750 M H₂SO₄ solution ?

Answer:

	Ca(OH)₂₍ₐq₎	+	H₂SO₄₍ₐq₎	→	CaSO₄ + H₂O
Reaction coefficients	1		1		
Volume (cm³)			52.68 (0.05268 L)		
No. of Moles	0.0395		0.0395		
Concentration (M)			0.75		
Mass	2.92				

Mr $[Ca(OH)_2]$ = 74

The number of moles in 52.68 cm³ of 0.750 M H_2SO_4 = volume x concentration = 52.88/1000 x 0.75 = 0.0395

According to the reaction coefficients, at the end point, the number of moles of $Ca(OH)_2$ = the number of moles of H_2SO_4

Thus, the number of moles of $Ca(OH)_2$ = 0.0395

Mass of 0.0395 moles $Ca(OH)_2$ = number of moles x M_r = 0.0395 x 74 = **2.92 g**

g) What volume of 1.0M sulfuric acid will be needed to neutralize 25.00 cm of 0.8M sodium hydroxide solution?

Answer:

	$2NaOH_{(aq)}$	+	$H_2SO_{4(aq)}$	→	$Na_2SO_4 + H_2O$
Reaction Coefficients	2		1		
Volume (cm³)	25 (0.025 L)		10 (0.01 L)		
No. of Moles	0.02		0.01		
Concentration (M)	0.8		1		

Number of moles in 25 cm³ of 0.8M NaOH = volume x concentration = 25/1000 x 0.8 = 0.02

At the end point, the number of moles of H_2SO_4 = 0.5 x number of moles of NaOH = 0.5 x 0.02 = 0.01

Volume of H_2SO_4 = number of moles /concentration = 0.01 / 1 = 0.01 L = **10 cm³**

h) Calculate the volume of 0.10 M sodium hydroxide solution needed to neutralize 20.00 cm³ of 0.25 M hydrochloric acid.

Answer:

	$NaOH_{(aq)}$	+	$HCl_{(aq)}$	→	$NaCl + H_2O$
Reaction coefficients	1		1		
Volume (cm³)	50 (0.05 L)		20 (0.02 L)		
No. of Moles	5×10^{-3}		5×10^{-3}		
Concentration (M)	0.1		0.25		

The number of moles in 20cm³ (0.02 L) of 0.32M HCl = volume x concentration = 20/1000 x 0.25 = 5×10^{-3}

At the end point, the number of moles of NaOH = number of moles of HCl = 5×10^{-3}

Volume of NaOH = number of moles/ concentration = 5×10^{-3} /0.1 = 0.05 L = **50cm³**

i) 100 cm³ of a magnesium hydroxide solution required 4.5 cm³ of 0.10 M sulfuric acid for complete neutralisation. Calculate the concentration of the magnesium hydroxide in grams per litre.

Answer:

	$Mg(OH)_{2(aq)}$	+	$H_2SO_{4(aq)}$	→	$MgSO_4 + H_2O$
Reaction Coefficients	1		1		
Volume (cm³)	100 (0.1 L)		4.5 (4.5 x 10⁻³ L)		
No. of Moles	4.5 x 10₋₅		4.5 x 10₋₄		
Concentration (M)	**4.5 x 10⁻³**		0.1		

Mr [Mg(OH)₂] = 58

Number of moles in 4.5 cm³ (4.5 x 10⁻³ L) of 0.1 M H₂SO₄ = volume x concentration = 4.5/1000 x 0.1 = 4.5 x 10⁻⁴

At the end point, the number of moles of H₂SO₄ = Number of moles of Mg(OH)₂

Thus the number of moles of Mg(OH)₂ in 100 cm³ (0.1 L) = 4.5 x 10⁻⁴

Concentration of Mg(OH)₂ = number of moles / volume = 4.5 x 10⁻⁵ /0.1
= **4.5 x 10⁻³ M**

To convert from M to g/mol multiply by M$_r$

Concentration of Mg(OH)₂ = 4.5 x 10⁻³ x 58 = **0.26 g/L**

j) Titration of 25.0 cm³ of a saturated calcium hydroxide solution required 35.45 cm³ of a 0.24 M hydrochloric acid solution for complete neutralisation. Calculate the molarity of the calcium hydroxide.

Answer:

	$Ca(OH)_{2(aq)}$	+	$2HCl_{(aq)}$	→	$CaCl_2 + 2H_2O$
Reaction coefficients	1		2		
Volume (cm³)	25 (0.025L)		35.45 (0.03545L)		
No. of Moles	8.51 x 10⁻³		8.51 x 10⁻³		
Concentration (M)	**0.17**		0.24		

At the end point of the titration, number of moles of Ca(OH)₂ = 1/2 x the number of moles of HCl

Thus number of moles in 35.45 cm³ (0.0345 L) of 0.24 M HCl	$= 35.45/1000 \times 0.24 = 8.51 \times 10^{-3}$
At the end point, the number of moles of $Ca(OH)_2$ in 25 cm³	$= 1/2 \times$ number of moles of HCl
	$= 1/2 \times 8.51 \times 10\text{-}3 = 4.254 \times 10^{-3}$
Concentration of $Ca(OH)_2$ = number of moles/volume	$= 4.254 \times 10^{-3} /0.025 = \mathbf{0.17\ M}$

Exercise 5.2

a) If it takes 50 cm³ of 0.5 M KOH solution to completely neutralize 125 cm³ of sulfuric acid solution (H_2SO_4), what is the concentration of the H_2SO_4 solution?

Answer:

	$2KOH_{(aq)}$	+	$H_2SO_{4(aq)}$	$\rightarrow$	$K_2SO_4 + 2H_2O$
Reaction coefficients	2		1		
Volume (cm³)	50		125		
No. of Moles	0.025		0.0125		
Concentration (M)	0.5		**0.1**		

The number of moles in 50cm³ of 0.5M KOH = volume x concentration = 50/1000 x 0.5 = 0.025

At the end point, the number of moles of H_2SO_4 = 1/2 x number of moles of KOH

Number of moles of KOH in 125 cm³ (0.125) = 1/2 x 0.025 = 0.0125

Concentration of KOH = number of moles/volume = 0.0125/0.125 = **0.1M**

b) Using phenolphthalein indicator, titration of 25.0 cm³ of 0.250M sodium hydroxide required 22.5 cm³ of a hydrochloric acid solution for complete neutralisation. What is the concentration of HCl?

Answer:

	$NaOH_{(aq)}$	+	$HCl_{(aq)}$	$\rightarrow$	$2NaCl + CO_2 + H_2O$
Reaction Coefficients	1		1		
Volume (cm³)	25		22.5		
No. of Moles	6.25×10^{-3}		6.25×10^{-3}		

Table contd.....

	NaOH$_{(aq)}$	+	HCl$_{(aq)}$	→	2NaCl + CO$_2$ + H$_2$O
Concentration (M)	0.25		**0.28**		

Number of moles in 25 cm^3 of 0.25 M NaOH = 25/1000 x 0.25 = 6.25 x 10^{-3}

At the end point the number of moles of NaOH in 25 cm^3 = number of moles of HCl in 22.5cm^3

Therefore, number of moles of HCl in 22.5 cm^3 (0.0225 L) = 6.25 x 10^{-3}

Concentration of HCl = number of moles/volume = 6.25 x 10^{-3}/0.0225 = 0.28 M

c) A solution made from pure barium hydroxide contained 2.74 g in exactly 100 cm^3 of water. Using phenolphthalein indicator, titration of 20.0 cm^3 of this solution required 18.7 cm^3 of a hydrochloric acid solution for complete neutralisation. What is the concentration of the acid?

Answer:

	Ba(OH)$_{2(aq)}$	+	2HCl$_{(aq)}$	→	BaCl$_2$ + H$_2$O
Reaction coefficients	1		2		
Volume (cm^3)	20 (0.02 L)		18.7 (0.0187 L)		
No. of Moles	0.0032		0.064		
Concentration (M)	0.16		**0.34**		

M$_r$ [Ba(OH)$_2$] = 171

Number of moles in 2.74 g of Ba(OH)$_2$ = mass/M$_r$ = 2.74/171 = 0.016

Number of moles of Ba(OH)$_2$ in 100 cm$_3$ (0.1 L) = 0.016

Concentration of Ba(OH)$_2$ = number of moles/ volume = 0.016 /0.1 = 0.16 M

Number of moles in 20 cm^3 of 0.16M Ba(OH)$_2$ = volume x concentration = 0.02 x 0.16 = 0.0032

At the end point the number of moles of HCl in 18.7 cm^3 = 2 x number of moles of Ba(OH)$_2$ in 20 cm^3

Therefore, number of moles of HCl in 18.7 cm^3 (0.0187) = 2 x 00032 = 0.0064

Concentration of HCl = number of moles /volume = 0.064/0.0187 = **0.34 M**

d) A 100 cm³ sample of water is titrated against 0.0334 M Na₄EDTA solution, and requires 8.84 cm³ to reach the end point. Calculate the concentration of CaCO₃ in the water sample, given the following equation

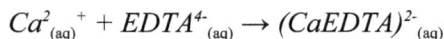

$$Ca^{2+}_{(aq)} + EDTA^{4-}_{(aq)} \rightarrow (CaEDTA)^{2-}_{(aq)}$$

Answer:

	$Ca^{2+}_{(aq)}$	+	$EDTA^{4-}$(aq)	→	$CaEDTA)^{2-}_{(aq)}$
Reaction Coefficients	1		1		
Volume (cm³)	100		8.4 (8.4 x 10⁻³ L)		
No. of Moles	2.8 x 10⁻⁴		2.8 x 10⁻⁴		
Concentration (M)	**2.8 x 10⁻⁴**		0.0334		

Number of moles in 8.4 cm³ of 0.0334 M EDTA⁴⁻ = 8.4/1000 x 0.0334 = 2.8 x 10⁻⁴

At the end point the number of moles of EDTA⁴⁻ in 8.4 cm³ = number of moles of Ca²⁺ in 100 cm³

Therefore, number of moles of Ca²⁺ (CaCO₃) in 100 cm³ (0.1 L)= 2.8 x 10⁻⁴

Concentration of CaCO₃ = number of moles /volume = 2.8 x 10-4/ 0.1 = **2.8 x 10-3 M**

e) What volume of 0.02 M KMnO₄ solution will react with 20cm³ of 0.1 M Fe²⁺(aq)? The reaction is:-

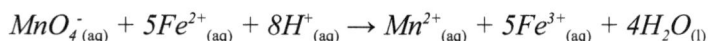

$$MnO_{4(aq)}^{-} + 5Fe^{2+}_{(aq)} + 8H^{+}_{(aq)} \rightarrow Mn^{2+}_{(aq)} + 5Fe^{3+}_{(aq)} + 4H_2O_{(l)}$$

Answer:

	$MnO_{4(aq)}$	+	$5Fe^{2+}_{(aq)}$	+ 8H⁺(aq)	→	Mn²⁺ + 5Fe³⁺ + 4H₂O
Reaction coefficients	1		5	8		
Volume (cm³)	20 (0.02 L)		20			
No. of Moles	4 x 10⁻⁴		0.002			

Table contd.....

	$MnO_{4\ (aq)}$	+	$5Fe^{2+}_{\ (aq)}$	$+ 8H^+_{\ (aq)}$	$\rightarrow$	$Mn^{2+} + 5Fe^{3+} + 4H_2O$
Concentration (M)	0.02		0.1			

Number of moles in 20 cm^3 of 0.1 M Fe^{2+} = 20/1000 x 0.2 = 0.002 moles

According to reaction coefficients tells us that 5 moles Fe^{2+} reacts with 1 mole of MnO_{4-}

Therefore, 0.002 moles of Fe^{2+} reacts with 0.002/5 = 0.0004 moles of MnO_{4-}

Volume of 0.02 M MnO_{4-} containing 0.0004 moles = 0.004/ 0.02 = 0.02 L = **20 cm^3**

f) 1.75 g of a hydrated sodium carbonate ($Na_2CO_3.xH_2O$) sample was dissolved in water and the volume made up to 250 cm^3. 25.0 cm^3 of this solution was found on titration, to require 24,5 cm^3 of 0. 1M HCl according the following equation.

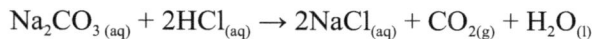

$$Na_2CO_{3\ (aq)} + 2HCl_{(aq)} \rightarrow 2NaCl_{(aq)} + CO_{2(g)} + H_2O_{(l)}$$

Calculate the value of x given the equation.

Answer:

	$Na_2CO_{3(aq)}$	+	$2HCl_{(aq)}$	$\rightarrow$	$2NaCl + CO_2 + H_2O$
Reaction coefficients	1		2		
Volume (cm^3)	25 (0.25 L)		24.5		
No. of Moles	1.225×10^{-3}		2.45×10^{-3}		
Concentration (M)			0.1		

Number of moles in 24.3 cm^3 of 0.1M HCl = volume x concentration = 24.5/1000 x 0.1 = 2.45×10^{-3}

At the end point, the number of moles of Na_2CO_3 in 25 cm^3 = 0.5 x number of moles of HCl in 24.5 cm^3 = 1.23×10^{-3}

Thus, number of moles of Na_2CO_3 in 250 cm^3 (0.25 L) = 250/25 x 1.23 x 10^{-3} = 0.0123

M_r [$Na_2CO_3.xH_2O$] = mass / number of moles = 1.75 / 0.0123 = 142

M_r[Na_2CO_3] = 106

$(H_2O)_n$ = 142 – 106 = 36, *i.e.,* **n = 2**

g) 25 cm³ of a sample of vinegar (CH₃COOH) was pipetted into a volumetric flask and the volume was made up to 250 cm³. This solution was placed in a burette and 13.9 cm³ were required to neutralize 25 cm³ of 0.1M NaOH. Calculate the molarity of the original vinegar solution and its concentration in g/L.

Answer:

	$CH_3COOH_{(aq)}$ +		$NaOH_{(aq)}$		→	$CH_3COONa + H_2O$
Reaction Coefficients	1		1			
Volume (cm³)	13.9 (0.0139 L)		25 (0.025 L)			
No. of Moles	2.5×10^{-3}		$25/1000 \times 0.1 = 2.5 \times 10^{-3}$			
Concentration (M)	0.18		0.1			

Number of moles in 25cm³ of 0.1M NaOH = $2.5/1000 \times 0.1 = 2.5 \times 10^{-3}$

According to the reaction coefficients, at the end point, number of moles of CH_3COOH in 13.9 cm³ = number of moles of NaOH in 25cm³

Number of moles of CH_3COOH in 13.9 cm³ = 2.5×10^{-3}

Thus, the number of moles of CH_3COOH in 250 cm³ = $250/13.9 \times 2.5 \times 10^{-3} = 0.045$

Concentration of diluted vinegar solution = number of moles / volume = 0.045/0.25 = 0.18 M

Therfore, the number of moles of CH_3COOH in the original undiluted 25 cm³ sample of vinegar (0.025 L) = 0.045

Concentration of original undilluted vinegar solution = number of moles / volume = 0.045 / 0.025 = **1.8M**

Mr[CH_3COOH] = 60

Therefore, concentration in g/L = Molarity x Mr = 1.8 x 60 = **108 g/L**

h) On titration, 0.2640 g of sodium oxalate dissolved in a flask required 30.74 cm³ of potassium permanganate to reach the end point. The equation for this reaction is:

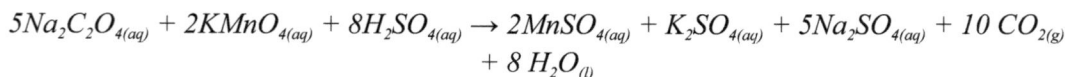

$$5Na_2C_2O_{4(aq)} + 2KMnO_{4(aq)} + 8H_2SO_{4(aq)} \rightarrow 2MnSO_{4(aq)} + K_2SO_{4(aq)} + 5Na_2SO_{4(aq)} + 10\ CO_{2(g)} + 8\ H_2O_{(l)}$$

i. *Determine the number of moles of oxalate in the flask.*

ii. *How many moles of potassium permanganate were required to reach the end point?*

iii. *Calculate the concentration of the potassium permanganate?*

Answer:

	$5Na_2C_2O_4$ +	$2KMnO_4$ +	$8H_2SO_4$ →	$2MnSO_4 + K_2SO_4 + 5Na_2SO_4 + 10$ $CO_2 + 8 H_2O$
Reaction coefficients	5	2	8	
Volume (cm³)		30.74		
		(0.03074 L)		
No. of Moles	1.97×10^{-3}	7.88×10^{-4}		
Concentration (M)		0.02567		

$M_r Na_2C_2O_4 = 134$

i. Number of moles in 0.2640 g of sodium oxalate = mass/M_r = $0.2640/134 = \mathbf{1.97 \times 10^{-3}}$

ii. According the reaction coefficients, 1 mole of $Na_2C_2O_4$ reacts with 2/5 moles of $KMnO_4$

Therefore, 1.97×10^{-3} moles of sodium oxalate will react with 2/5 x 1.97×10^{-3} moles of $KMnO_4 = \mathbf{7.88 \times 10^{-4}}$

iii. Concentration of $KMnO_4$ = number of moles/ volume = 7.88×10^{-4} /0.0374 = **0.02567M**

i) *A 2.5 g sample of ethanedioic acid, $H_2C_2O_4.nH_2O$, was dissolved in water and the solution made up to 250 cm³. 15.8 cm³ of this solution required to 25 cm³ of 0.1 M NaOH for complete neutralization. Calculate the value of n given that ethanedioic acid reacts with NaOH in a 1:2 ratio.*

Answer:

	$H_2C_2O_4.nH_2O_{(aq)}$ +	$2NaOH^{(aq)}$ →	$2NaC_2O_4 + (2 +n)H_2O$
Reaction Coefficients	1	2	
Volume (cm³)	15.8	25	

Table contd.....

	$H_2C_2O_4.nH_2O_{(aq)}$	+	$2NaOH^{(aq)}$	→	$2NaC_2O_4 + (2 +n)H_2O$
No. of Moles	1.25×10^{-3}		2.5×10^{-3}		
Concentration (M)	0.79		0.1		

Number of moles in 25cm^3 of 0.1M NaOH = volume x concentration = 25/1000 x 0.1 = 2.5×10^{-3}

At the end point, number of moles $H_2C_2O_4.nH_2O$ in 15.8 cm^3 (0.0158) = 0.5 x number of moles of NaOH = 0.5 x 2.5×10^{-3} = 1.25×10^{-3}

Thus, number of moles of $H_2C_2O_4.nH_2O$ in 250 cm^3 = 250/15.8 x 1.25 x 10^{-3} = 0.0198

$M_rH_2C_2O_4.nH_2O$ = 1/0.0198 x 2.5 = 126

$M_rH_2C_2O_4$ = 90

Thus M_rH_2On = 126 – 90 = 36

Therefore, since $M_r H_2On$ = 18, **n= 2**

j) A 3.50 g sample of lawn sand containing an iron (II) salt was shaken with dilute H_2SO_4. This solution required 25.00 cm^3 of 0.0200 M potassium manganate (VII) for complete oxidation of the Fe (II) ions in the solution back to Fe(III) ions. Using the equation given below, determine the percentage of Fe(II) ions (by mass) in the lawn sand sample.

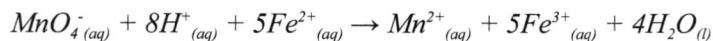

$$MnO_{4\ (aq)}^- + 8H^+_{\ (aq)} + 5Fe^{2+}_{\ (aq)} \rightarrow Mn^{2+}_{\ (aq)} + 5Fe^{3+}_{\ (aq)} + 4H_2O_{(l)}$$

Answer:

	MnO_4^-	+	$8H^+$	+	$5Fe^{2+}$	→	$Mn^{2+} + 5Fe^{3+} + 4H_2O$
Reaction Coefficients	1		8		5		
Volume (cm^3)	25 (0.025 L)						
No. of Moles	5×10^{-4}				2.5×10^{-3}		
Mass(g)					**0.14**		
Concentration (M)	0.02						

Number of moles in 25 cm^3, 0.02M $KMnO_4$ = 25/1000 x 0.02 = 5×10^{-4}

According to the reaction coefficients the number of moles of Fe^{2+} = 5 x number of moles of $KMnO_4$ = 5 x 5×10^{-4} = 2.5×10^{-3}

Mass of Fe = number of moles x Mr = $2.5 \times 10^{-3} \times 56 = 0.14$ g

% composition of Fe = mass of Fe/ mass of sample = 0.14/3.5 x 100 =
4%

CHAPTER 6

Molar Volumes & Reacting Volumes

Keywords: Calculations, Gases, Molar volume, Moles to volume, Reacting volumes.

6.1. MOLES & GASES

At Standard Temperature and Pressure (STP, 0°C, 1 atmosphere pressure), one mole of gas occupies a volume of 22.4 litres (22.4 L; 22,400 cm³). Thus at STP, 28g of nitrogen gas (N_2) occupies a volume of 22.4 litres, 32 g of oxygen (O_2) gas occupies a volume of 22.4 litres and 71 g of chlorine gas (Cl_2) occupies a volume of 22.4 litres. Similarly, 22.4 litres of hydrogen gas has a mass of 2 g and 22.4 litres of helium gas has a mass of 4 g at STP.

In other words, a given volume of gas always contains the same number of formula units (particles) regardless of the substance at a given temperature and pressure. But, the mass of one mole of gas (M_r in grams) differs from substance to substance.

ONE MOLE = M_r in grams = 6.02 x 10^{23} particles = 22.4 litres of a gas at STP

Given that the relative formula mass of CO_2 = 44 (12 +16 +16):

44g of CO_2 contains 6 x 10^{23} molecules of CO_2 which equals 1 mole and under STP occupies a volume of 22,400 cm³ (22.4 L)

Thus,

0.5 moles of CO_2 has a mass of 22g, contains 3 x10^{23} molecules of CO_2 and occupies a volume of 11,200 cm³ (11.2 litres) at STP

Equal volumes of gases at the same temperature and pressure contain the same number of moles.

Nigel P. Freestone
All rights reserved-© 2016 Bentham Science Publishers

The molar volume (V_m) is simply defined as the **volume occupied by one mole of a gas**. The units of molar volume are L/mol (dm^3/mol or cm^3/mole).

At STP one mole of a gas occupies a volume of 22.4 L (22,400 cm^3). Molar volume at Room Temperature and Pressure (RTP, 25°C, 1 atmosphere) is 24.0 L (24,000 cm^3).

Molar Volume varies with both temperature and pressure.

Volume of gas = Number of moles x V_m
Number of moles = Volume of gas / V_m
V_m = Volume of Gas/Number of moles

Ensure that the volume units for V and V_m are the same, *i.e.* either L (dm^3) or cm^3.

6.2. MOLES TO VOLUME

Volume of gas = Number of moles x V_m

Example 6.1: *If the molar volume is 24 L/mol, what is the volume is occupied by 0.025 mole of H_2?*

Answer:

1 mole occupies a volume of 24 L
Volume of gas = Number of moles x V_m
Thus 0.25 moles will occupy 0.025 x 24 = 0.6 L = **600 cm^3**

Example 6.2: *Determine the volume occupied by 0.5 moles of chlorine gas given that the molar volume is 23,000 cm^3/mol.*

Answer:

V_m = 23,000 cm^3
Number of moles = 0.5
Volume of gas = Number of moles x V_m
Thus 0.5 moles will occupy 0.5 x 23000 = 11,500 cm^3 = **11.5 L**

Exercise 6.1

Calculate the volume occupied by the following gases at STP (V_m = 22.4 litres)

a. *2 mole of CH_4*
b. *0.3 moles of NH_3*
c. *1.6 moles of C_2H_4*
d. *3 moles of SO_2*
e. *0.26 moles of NO*
f. *5.7 moles of HBr*
g. *0.22 moles of Cl_2*
h. *0.020 moles of CO_2*
i. *15 moles of O_2*
j. *3.5 x 10^{-3} moles of H_2*

6.3. VOLUME TO MOLES (AND MASS)

$$\text{Number of moles} = \text{Volume} / V_m$$
$$\text{Mass} = \text{Number of moles} \times M_r$$

Example 6.3: *Calculate the number of moles of carbon dioxide in 250 cm^3 of the gas. Assume that one mole of a gas occupies a volume of 23,000 cm^3*

Answer:

V_m = 23 L = 23,000 cm^3
Number of moles = Volume / V_m
Thus number of moles present in 250 cm^3 = 250/23000 = **0.011 moles**

Example 6.4: *What is the mass 5 litres of oxygen gas at STP?*

Answer:

V_m (at STP) = 22.4 L = 22,400 cm^3
$M_r[O_2]$ = 32
Number of moles = Volume of gas / V_m
Number of moles present in 5 L = 5/22.4 = 0.223 moles
Mass of O_2 = number of moles x M_r

Mass of O_2 = 0.223 x 32 = **7.14 g**

Exercise 6.2

Determine the number of moles present in the following volumes of gases at room temperature and pressure. Assume one mole of gas occupies a volume of 24 dm³ (24 000 cm³) under these conditions.

a. *200 cm³ of CH_4*
b. *10 cm³ of NH_3*
c. *3400 cm³ of C_2H_4*
d. *657 cm³ of SO_2*
e. *400 cm³ of NO*
f. *2400 cm³ of HBr*
g. *700 L of Cl_2*
h. *0.5 L of CO_2*
i. *9.5 x 10⁻³L of O_2*
j. *7.89 L of H_2*

Exercise 6.3

Calculate the mass of each gas. Assume that 1 mole of gas occupies a volume of 24 L.

a. *200 cm³ of CH_4*
b. *10 cm³ of NH_3*
c. *3400 cm³ of C_2H_4*
d. *657 cm³ of SO_2*
e. *1400 cm³ of NO*
f. *2400 cm³ of HBr*
g. *760 cm³ of Cl_2*
h. *5 L of CO_2*
i. *9.5 L of O_2*
j. *7.89 L of H_2*

6.4. RELATIVE MOLECULAR MASS FROM VOLUME & MASS

$$\text{Number of moles} = \text{Volume} /V_m$$
$$M_r = \text{Mass/Number of moles}$$

Given the relationship between mass, the number of moles and the volume occupied by a gas, the relative formula mass (M_r) of a gas can easily be determined from the volume occupied by a known mass of gas under at a given set of temperature and pressure conditions.

Example 6.5: *100 cm³ of a gas at STP has a mass of 0.1964 g. Calculate the Relative Molecular Mass of the gas.*

Answer:

This requires us to find the volume occupied by 1 mole of the gas, *i.e.*, 22,400 cm³ (22.4 L).
Number of moles of the gas in 0.0667g = 100/22400 = 0.004464 moles.
Mr = Mass/Number of moles = 0.1964/0.004464 = **44 g mol⁻¹**

Exercise 6.4

Calculate the relative molecular mass of each gas from the given data. Assume that all volumes are measured at STP ($V_m = 22,400$ cm³).

a. *0.442 g of gas occupy 66 cm³*
b. *0.714 g of gas occupy 250 cm³*
c. *0.402 g of gas occupy 100 cm³*
d. *2.25 g of gas occupy 600 cm³*
e. *0.0893 g of gas occupy 1000 cm³*
f. *0.1875 g of gas occupy 150 cm³*
g. *1.90 g of gas occupy 600 cm³*
h. *0.235 g of gas occupy 94 cm³*
i. *4.286 g of gas occupy 1200 cm³*
j. *2.143 g of gas occupy 1600 cm³*

6.5. MOLAR VOLUME FROM MASS

$$V_m = \text{Volume of gas/ Number of moles}$$
$$\text{Mass} = M_r \text{ x Number of moles}$$

Example 6.6: *976 cm³ of oxygen was found to have a mass of 1.3 g. Calculate the molar volume of oxygen, under these conditions.*

Answer:

V_m = Volume of gas/ Number of moles
Volume of gas = 976 cm³
$M_r[O_2] = 32$
Number of moles in 1.3 g of O_2 = mass/M_r = 1.3/32 = 0.0406
V_m = 976/0.0406 = 24,039 cm³ = **24.04 L**

6.6. VOLUME OF ONE GAS, FROM ANOTHER GAS

The following examples illustrate how straightforward it is to calculate the volume of a product gas, from the volume of a reactant gas, and *vice versa*.

Example 6.7: *Calculate the volume propane, C_3H_8, which must be combusted to generate 100 cm³ of carbon dioxide gas (under the same conditions of temperature and pressure).*

Answer:

Write a balanced chemical equation for the reaction and construct a mole calculation frame around it.

	C_3H_8	+	$5O_2$		→	$3CO_2$	+	$4H_2O$
Volume (cm₃)	100/3 = **33.3**		5/3 x 100 = 167			100		4/3 x 100 = 133
Reaction coefficients	1		5			3		4

Equal volumes of gases at the same temperature and pressure contain the same number of moles.
Thus, according to the reaction coefficients in the balanced equation, 1 volume of

C_3H_8 will be totally combusted by 5 volumes of O_2 to produce 3 volumes of CO_2 and 4 volumes of H_2O.

Therefore 100 cm³ CO_2 will be produced from combustion of 100/3 = **33.3 cm³ of C_3H_8**

Example 6.8: *Write a balanced equation for the formation of water from the reaction between hydrogen and oxygen, and then calculate the volume of oxygen needed to react with 123 cm³ of hydrogen.*

Answer:

	2H$_2$	**+**	**O$_{2(g)}$**	**→**	**2H$_2$O$_{(l)}$**
Volume (cm₃)	123		62.5		
Reaction coefficients	2		1		2

Equal volumes of gases at the same temperature and pressure contain the same number of moles.

According to the reaction coefficients in the balanced equation, 2 volumes of H_2 react with 1 volume of O_2 to produce 2 volumes of H_2O.

Therefore, 123 cm³ of H_2 requires 123/2 = **62.5 cm³ of O_2** for complete combustion.

6.7. EQUATIONS, MOLES AND VOLUMES

Example 6.9: *Calculate volume of hydrogen required to produce 8.2 g of cyclohexane from cyclohexene (V_m = 24 litres)*

$$C_6H_{10(l)} + H_{2(g)} \rightarrow C_6H_{12(g)}$$

Answer:

	C$_6$H$_{10(l)}$	**+**	**H$_{2(g)}$**	**→**	**C$_6$H$_{12(l)}$**
A$_r$ /M$_r$	82		2		84
Mass	8.2				
No. of moles	8.2/82 = 0.1				
Volume (L)	0.1 x 24 = 2.4				
Reaction coefficients	1		1		1

$M_r [C_6H_{10}] = (6 \times 12) + (10 \times 1) = 82$

Number of moles in 8.2 g of $C_6H_{10} = 8.2/82 = 0.1$ moles

According to the reaction coefficients in the balanced chemical equation, 1 mole of H_2 can hydrogenate 1 mole C_6H_{10}.

Thus, 0.1 moles of H_2 are required to hydrogenate 0.1 moles of C_6H_{10}.

Volume occupied by 0.1 moles of H_2 = number of moles x V_m = 0.1 x 24 = **2.4 L** = **2,400 cm³**

Example 6.10: *Calculate the volume of hydrogen generated at room temperature and pressure (RTP) when 12 g of magnesium reacts with an excess of sulfuric acid.*

Answer:

	$2Mg_{(s)}$	+	$H_{2(g)}SO_{4(l)}$	→	$MgSO_{4(g)}$	$H_{2(g)}$
A_r/M_r	24		98		120	2
Mass	12					0.5
No. of moles						0.25 x 24 = 6
Volume (L)	0.5		0.25		0.25	0.5/2 = 0.25
Reaction coefficients	2		1		1	1

V_m (at RTP) = 24 L.

$A_r [Mg] = 24$ and $M_r [H_2] = 2$.

Number of moles in 12 g of Mg = mass/M_r = 12/24 = 0.5.

According to the reaction coefficients in the balanced chemical equation, 2 moles of Mg produces 1 mole of H_2.

Thus 0.5 moles of Mg can produce 0/5/2 = 0.25 moles of H_2.

Volume occupied by 0.25 moles H_2 = number of moles x V_m = 0.25 x 24 = 6 L.

Volume of H_2 generated = **6L**

Exercise 6.5

a) Calculate the volume of 0.25 mole of carbon dioxide, when the molar volume is 24.0 L mol⁻¹? Give your answer in litres.

b) Assuming that air is 20% oxygen by volume, determine the number of moles of oxygen present in 20.0 L of air at STP.

c) Calculate the molar volume if 0.25 moles of a gas occupies a volume of 4,680 cm³.

d) The complete combustion of methane produces carbon dioxide and water vapour:

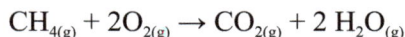

$$CH_{4(g)} + 2O_{2(g)} \rightarrow CO_{2(g)} + 2\,H_2O_{(g)}$$

Calculate the volume of oxygen required to completely combust 125 cm³ of methane and the volume of carbon dioxide that would be generated.

e) What volume of propane, C_3H_8, must be burned to generate 60 cm³ of carbon dioxide (under the same conditions of temperature and pressure)?

f) Determine the volume of ammonia produced when 20 L of hydrogen reacts with an excess of nitrogen (under the same conditions of temperature and pressure):

$$2N_{2(g)} + 3H_{2(g)} \rightarrow 2NH_{3(g)}$$

g) What volume of O_2 is consumed when 25cm³ of benzene gas is fully combusted (under the same conditions of temperature and pressure)?

h) What volume of nitrogen dioxide would be produced from the complete combustion of 3 litres of nitrogen gas (under the same conditions of temperature and pressure)?

i) The combustion of ammonia produces nitrogen monoxide:

$$4NH_{3(g)} + 5O_{2(g)} \rightarrow 4NO_{(g)} + 6H_2O_{(l)}$$

Calculate the volume of ammonia required to produce 10 L of nitrogen monoxide.

j) Sulphur trioxide used in the manufacture of sulphuric acid is produced from the oxidation of sulphur dioxide.

$$2SO_{2(g)} + O_{2(g)} \rightarrow SO_{3(g)}$$

Assuming 20% of the air by volume is oxygen, calculate the volume of air required to generate 200 cm³ of sulphur trioxide?

Exercise 6.6

a) Calculate the mass of magnesium required to generate 200 cm³ H$_2$ at STP.

$$Mg_{(s)} + 2HCl_{(aq)} \rightarrow MgCl_{2\,(aq)} + H_{2\,(g)}$$

b) Determine volume of ozone, at STP, produced when 0.95 g of fluorine is reacted with an excess of steam:

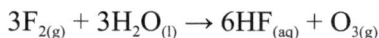

$$3F_{2(g)} + 3H_2O_{(l)} \rightarrow 6HF_{(aq)} + O_{3(g)}$$

c) Lead sulphide is converted to lead (II) oxide when heated in oxygen:

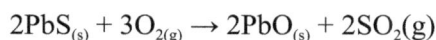

$$2PbS_{(s)} + 3O_{2(g)} \rightarrow 2PbO_{(s)} + 2SO_2(g)$$

Determine the volume of oxygen (molar volume = 24L) that would react with exactly 95.72 g of PbS?

d) The chemical weapon phosgene is produced industrially according to the following reaction:

$$CO_{(g)} + Cl_{2(g)} \rightarrow COCl_{2(g)}$$

Calculate the volume of CO (at STP) required to produce 10 g of phosgene.

e) What mass of ammonium chlorate is needed to decompose to give off 100 litres of oxygen at STP?

$$2KClO_{3(s)} \rightarrow 2\ KCl_{(s)} + 3O_2(g)$$

f) What volume of oxygen is required to completely combust a kilogram of octane at RTP?

g) How many grams of sodium do you have to put into water to make 30 litres of hydrogen at STP?

$$2Na_{(s)} + H_2O_{(l)} \rightarrow Na_2O_{(aq)} + H_{2(g)}$$

h) Calculate the volume of oxygen produced at STP from the decomposition of 1.7 g of hydrogen peroxide (H$_2$O$_2$).

$$2H_2O_{2(l)} \rightarrow 2H_2O_{(l)} + O_{2(g)}$$

i) Sulphur dioxide can be removed from power station flue gas by passing it through calcium hydroxide slurry. The equation for this reaction is:

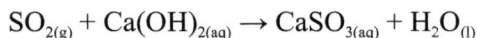

$$SO_{2(g)} + Ca(OH)_{2(aq)} \rightarrow CaSO_{3(aq)} + H_2O_{(l)}$$

What mass of calcium hydroxide would be needed to remove 1000 dm³ of sulphur dioxide at STP?

j) What volume of oxygen, at STP, is required to completely combust 90 g ethane?

ANSWERS

Exercise 6.1

Calculate the volume occupied by the following gases stated at STP (V_m = 22.4 litres).

a) 2 mole of CH_4

Answer:

Volume   = Number of moles x V_m
    = 2 x 22.4
    = **44.8 L**

b) 0.3 moles of NH_3

Answer:

Volume   = Number of moles x V_m
    = 0.3 x 22.4
    = **6.72 L**

c) *1.6 moles of C_2H_4*

Answer:

Volume = Number of moles x V_m
= 1.6 x 22.4
= **35.84 L**

d) *3 moles of SO_2*

Answer:

Volume = Number of moles x V_m
= 3 x 22.4
= **67.2 L**

e) *0.26 moles of NO*

Answer:

Volume = Number of moles x V_m
= 0.26 x 22.4
= **5.82 L**

f) *5.7 moles of HBr*

Answer:

Volume = Number of moles x V_m
= 5.7 x 22.4
= **127.68 L**

g) 0.22 moles of Cl_2

   Answer:

   Volume   = Number of moles x V_m
                   = 0.22 x 22.4
                   = **4.93 L**

h) 0.020 moles of CO_2

   Answer:

   Volume   = Number of moles x V_m
                   = 0.020 x 22.4
                   = **0.448 L**

i) 15 moles of O_2

   Answer:

   Volume   = Number of moles x V_m
                   = 15 x 22.4
                   = **336 L**

j) 3.5×10^{-3} moles of H_2

   Answer:

   Volume   = Number of moles x V_m
                   = 3.5×10^{-3} x 22.4
                   = **0.0784 L**

Exercise 6.2

Determine the number of moles present in the following volumes of gases. Assume that all volumes are measured at room temperature and pressure. Assume one mole of gas occupies a volume of 24 dm³ (24 000 cm³) under these conditions.

Answer:

It is essential that you use the same units for the volume of gas and V_m, *i.e.* either L or cm^3.

a) 200 cm^3 of CH_4

> *Answer:*
>
> Number of moles   = Volume /V_m
> = 200/24000
> = **8.33 x 10^{-3}**
>
> OR
>
> 200 cm^3 = 200/1000 = 0.2 L
> Number of moles   = 0.2/24
> = **8.33 x 10^{-3}**

b) 10 cm^3 of NH_3

> *Answer:*
>
> Number of moles   = Volume /V_m
> = 10/24000
> = **4.17 x 10^{-4}**

c) 3400 cm^3 of C_2H_4

> *Answer:*
>
> Number of moles   = Volume /V_m
> = 3400/24000
> = **0.142**

d) 657 cm³ of SO₂

Answer:

Number of moles   = Volume /V_m
                                = 657/24000
                                = **0.027**

e) 400 cm³ of NO

Answer:

Number of moles   = Volume /V_m
                                = 1400/24000
                                = **0.0167**

f) 2400 cm³ of HBr

Answer:

Number of moles   = Volume /V_m
                                = 2400/24000
                                = **0.1**

g) 700 L of Cl₂

Answer:

Number of moles   = Volume /V_m
                                = 700/24
                                = **29.1**

h) 0.5 L of CO_2

Answer:

Number of moles   = Volume /V_m
                        = 0.5/24
                        = **0.021**

i) 9.5 L of O_2

Answer:

Number of moles   = Volume /V_m
                        = 9.5/24
                        = **0.396**

j) 7.89 L of H_2

Answer:

Number of moles   = Volume /V_m
                        = 7.89/24
                        = **0.329**

Exercise 6.3

Calculate the mass of each gas. Assume that 1 mole of gas occupies a volume of 24 L.

a. 200 cm^3 of CH_4

Answer:

M_r [CH_4] = 16
Number of moles = Volume/V_m
Number of moles = 200/24000 = 8.33×10^{-3}
Mass = Number of moles x M_r

Mass = $8.33 \times 10^{-3} \times 16 =$ **0.13 g**

b. *10 cm³ of NH₃*

Answer:

M_r [NH_3] = 17
Number of moles = Volume/V_m
Number of moles = 10/24000 = 4.17×10^{-4}
Mass = $4.17 \times 10^{-4} \times 17 =$ **7.08 $\times 10^{-3}$ g**

c. *3400 cm³ of C₂H₄*

Answer:

M_r [C_2H_4] = 28
Number of moles = Volume/V_m
Number of moles = 3400/24000 = 0.142
Mass = Number of moles x M_r
Mass = 0.142 x 28 = **3.97 g**

d. *657 cm³ of SO₂*

Answer:

M_r [SO_2] = 64
Number of moles = Volume/V_m
Number of moles = 657/24000 = 0.0273
Mass = Number of moles x M_r
Mass = 0.0273 x 64 = **1.752 g**

e. *1400 cm³ of NO*

Answer:

M_r [NO] = 30

Number of moles = Volume/V_m
Number of moles = 1400/2400 = 0.058
Mass = Number of moles x M_r
Mass = **1.75 g**

f. *2400 cm³ of HBr*

Answer:

M_r [HBr] = 81
Number of moles = Volume/V_m
Number of moles = 2400/24000 = 0.1
Mass = Number of moles x M_r
Mass = 0.1 x 81 = **8.1 g**

g. *760 cm³ of Cl₂*

Answer:

M_r [Cl₂] = 71
Number of moles = Volume/V_m
Number of moles = 760/24000 = 0.0317
Mass = Number of moles x M_r
Mass =0.03127 x 71 = **2.25 g**

h. *5 L of CO₂*

Answer:

M_r [CO₂] = 44
Number of moles = Volume/V_m
Number of moles = 5/24 = 0.208 g
Mass = Number of moles x M_r
Mass =0.208 x 44 = **9.17 g**

i. *9.5 L of O_2*

Answer:

$M_r [O_2] = 32$
Number of moles = Volume/V_m
Number of moles = 9.5/24 = 0.396
Mass = Number of moles x M_r
Mass =0.396 x 32 = **12.7 g**

j. *7.89 L of H_2*

Answer:

$M_r [H_2] = 2$
Number of moles = Volume/V_m
Number of moles =7.89/24 = 0.329
Mass = Number of moles x M_r
Mass = 0.329 x 2 = **0.658 g**

Exercise 6.4

Calculate the relative molecular mass of each gas from the given data. Assume that all volumes are measured at STP ($V_m = 22,400$ cm³).

a. *0.442 g of gas occupy 66 cm³*

Answer:

Number of moles in 66 cm³ of a gas at STP = V/V_m = 66/22400 = 2.946 x 10^{-3}
Thus 2.946 x 10^{-3} moles has a mass of 0.442 g
M_r = mass / number of moles = 0.442/ 2.946 x 10^{-3} = **150**

b. *0.714 g of gas occupy 250 cm³*

   Answer:

   Number of moles in 250 cm³ of gas at STP = V/V_m = 250/22400 = 0.1116
   Thus 0.1116 moles has a mass of 0.714 g
   M_r = mass / number of moles = 0.714/ 0.1116 = **64**

c. *0.402 g of gas occupy 100 cm³*

   Answer:

   Number of moles in 100 cm³ of a gas = V/V_m = 100/22400 = 4.464 x 10^{-3}
   Thus 4.464 x 10^{-3} moles of gas has a mass of 0.402 g
   M_r = mass / number of moles = 0.402/.464 x 10^{-3} = **90**

d. *2.25 g of gas occupy 600 cm³*

   Answer:

   Number of moles of gas = V/V_m = 600/22400 = 0.0268
   Thus 0.0268 moles of gas has a mass of 2.25 g
   M_r = mass / number of moles = 2.25 4.464 x 10^{-3} =**84**

e. *0.0893 g of gas occupy 1000 cm³*

   Answer:

   Number of moles of gas = V/V_m = 1000/22400 = 0.0446
   Thus 0.0446 moles of gas has a mass of 0.0893 g
   M_r = mass / number of moles 0.0893/ 0.0446 = **2**

f. *0.1875 g of gas occupy 150 cm³*

   Answer:

   Number of moles of gas = V/V_m = 150/22400 = 6.7 x 10^{-3}

Thus 6.7×10^{-3} moles of gas has a mass of 0.1875 g

M_r = mass / number of moles = $0.1875 / 6.7 \times 10^{-3}$ = **28**

g. *1.90 g of gas occupy 600 cm³*

Answer:

Number of moles of gas = V/V_m = 600/22400 = 0.0268

Thus 0.0268 moles of gas has a mass of 1.90 g

M_r = mass / number of moles = 1.90 / 0.0268 x 1.90 = **71**

h. *0.235 g of gas occupy 94 cm³*

Answer:

Number of moles of gas = V/V_m = 94/22400 = 4.18×10^{-3}

Thus 4.18×10^{-3} moles of gas has a mass of 0.235 g

M_r = mass / number of moles =0.235 / 4.18×10^{-3} = **56**

i. *4.286 g of gas occupy 1200 cm³*

Answer:

Number of moles of gas = V/V_m = 1200/22400 = 0.0536

Thus 0.0536 moles of gas has a mass of 4.286 g

M_r = mass / number of moles = 4.286 /0.0536 = **80**

j. *2.143 g of gas occupy 1600 cm³*

Answer:

Number of moles of gas = V/V_m = 1600/22400 = 0.0714

Thus 0.0714 moles of gas has a mass of 2.143 g

M_r = mass / number of moles = 2.143 /0.0714 = **30**

Exercise 6.5

a) *Calculate the volume of 0.25 mole of carbon dioxide, when the molar volume is 24.0 L mol⁻¹? Give your answer in litres.*

   Answer:

   One mole of CO_2 occupies 24 L.
   Therefore, 0.25 moles CO_2 will occupy = number of moles x V_m = 0.25 x 24 = **6 L**

b) *Assuming that air is 20% oxygen by volume, determine the number of moles of oxygen is present in 20.0 L of air at STP?*

   Answer:

   Volume of O_2 in 20 L of air = 20/100 x 20 = 4 L.
   Number of moles of O_2 in 4 L = volume/V_m = 4/22.4 = **0.179**

c) *Calculate the molar volume if 0.25 moles of a gas occupies a volume of 4,680 cm³.*

   Answer:

   0.25 moles occupies a volume of 4680 cm³.
   V_m = Volume / Number of moles = 4680 / 0.25 = 18,720 cm³ = **18.72 L**

d) *The complete combustion of methane produces carbon dioxide and water vapour:*

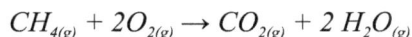

$$CH_{4(g)} + 2O_{2(g)} \rightarrow CO_{2(g)} + 2\,H_2O_{(g)}$$

   Calculate the volume of oxygen required to completely combust 125 cm³ of methane and the volume of carbon dioxide that would be generated.

Table contd.....

	$CH_{4(g)}$	+	$2O_{2(g)}$	$\rightarrow$	$CO_{2(g)}$	+	$2H_2O_{(g)}$
Reaction coefficients	1		2		1		2
Volume (cm³)	125		2 x 125 = 250		125		250

Equal volumes of gases at the same temperature and pressure contain the same number of moles.

According to the reaction coefficients in the balanced equation, 1 volume of CH_4 reacts with 2 volumes of O_2 to produce 1 volume of CO_2 and 2 volumes of H_2O.

Thus, 125 cm³ CH_4 requires 2 x 125 = **250 cm³** of **O_2** for total combustion, and the combustion of 125 cm³ of CH_4 will produce **125 cm³** of **CO_2**.

e) What volume of propane, C_3H_8, must be burned to generate 60 cm³ of carbon dioxide gas (under the same conditions of temperature and pressure)?

Answer:

	$C_3H_{8(l)}$	+	$5O_{2(g)}$	$\rightarrow$	$3CO_{2(g)}$	+	$4H_2O_{(g)}$
Reaction coefficients	1		5		3		4
Volume (cm³)	60/3 = **20**		60/3 x 5 = **100**		60		

Write a balance chemical equation and construct a calculating frame around it.

Equal volumes of gases at the same temperature and pressure contain the same number of moles.

According to the reaction coefficients in the balanced equation, 1 volume of C_3H_8 reacts with 5 volumes of O_2 to produce 3 volumes of CO_2 and 4 volumes of H_2O.

Therefore, 60 cm³ CO_2 will be produced from the combustion of 60/3 = **20 cm³ of C_3H_8**

f) Determine the volume of ammonia produced when 20 L of hydrogen reacts with an excess of nitrogen (under the same conditions of temperature and pressure):

$$2N_{2(g)} + 3H_{2(g)} \rightarrow 2NH_{3(g)}$$

Answer:

	$2N_{2(l)}$	+	$3H_2$	$\rightarrow$	$2NH_{3(g)}$
Reaction coefficients	2		3		2
Volume (cm³)	20/3 x 2 = 13.33		20		20/3 x 2= **13.33**

Equal volumes of gases at the same temperature and pressure contain the same number of moles.

According to the reaction coefficients in the balanced equation, 2 volumes of N_2 reacts with 3 volumes of H_2 to produce 2 volumes of NH_3.

Therefore, 20 L of N_2 will produce 20 x 3/2 = **13.33 L of NH₃**

g) *What volume of O_2 is consumed when 25cm³ of benzene is fully combusted oxygen (under the same conditions of temperature and pressure)?*

Answer:

	$2C_6H_{6(l)}$	+	$15O_{2(g)}$	$\rightarrow$	$12CO_{2(g)}$	+	$6H_2O_{(g)}$
Reaction coefficients	2		15		12		6
Volume (cm³)	25		25 x 15/2 = **187.5**		25 x 12/2 = 150		25 x 6/2 = 75

Equal volumes of gases at the same temperature and pressure contain the same number of moles.

According to the reaction coefficients in the balanced equation, 2 volumes of C_6H_6 react with 15 volumes of O_2 to produce 3 volumes of CO_2 and 4 volumes of H_2O.

Therefore, the volume of O_2 required to combust 25 cm³ of C_6H_6 = (15/2) x 25 = **187.5 cm³**

h) *What volume of nitrogen dioxide would be produced from the complete combustion of 3 litres of nitrogen gas (under the same conditions of temperature and pressure)?*

Answer:

	$N_{2(g)}$	+	$2O_{2(g)}$	$\rightarrow$	$2NO_{2(g)}$
Reaction coefficients	1		2		2
Volume (cm^3)	3		6		6

Equal volumes of gases at the same temperature and pressure contain the same number of moles.

According to the reaction coefficients in the balanced equation, 1 volume of N_2 reacts with 2 volumes of O_2 to produce 2 volumes of NO_2.

Therefore, total combustion of 3 L N_2 will produce 3 x 2 = **6 L of NO_2**

i) The combustion of ammonia produces nitrogen monoxide:

$$4NH_{3(g)} + 5O_{2(g)} \rightarrow 4NO_{(g)} + 6H_2O_{(l)}$$

Calculate the volume of ammonia required to produce 10 L of nitrogen monoxide and determine the volume of air would be used in the conversion.

Answer:

	$4NH_{3(g)}$	+	$5O_{2(g)}$	$\rightarrow$	$4NO_{(g)}$	+	$6H_2O_{(g)}$
Reaction coefficients	4		5		4		6
A_r /M_r	17		32		30		18
Volume (L)	**10**		12.5		10		15

Equal volumes of gases at the same temperature and pressure contain the same number of moles.

According to the reaction coefficients in the balanced equation, 4 volumes of NH_3 react with 5 volumes of O_2 to produce 4 volumes of NO and 6 volumes of H_2O.

Therefore, the number volume of NH_3 required to generate 10 L of NO = **10 L**

j) Sulphur trioxide used in the manufacture of sulphuric acid is produced from the oxidation of sulphur dioxide.

$$2SO_{2(g)} + O_{2(g)} \rightarrow SO_{3(g)}$$

What volume of air (assume 20% of the air is oxygen) would be needed to produce 200 cm³ of sulphur trioxide? Assume complete conversion of sulphur dioxide to sulphur trioxide.

Answer:

	$2SO_{2(aq)}$	+	$O_{2(aq)}$	→	$2SO_{3(aq)}$
Reaction coefficients	2		1		2
Volume (cm³)	200		100		200

Equal volumes of gases at the same temperature and pressure contain the same number of moles.

According to the reaction coefficients in the balanced equation, 2 volumes of SO_2 react with 1 volume of O_2 to produce 2 volumes of SO_3. Therefore, the volume of O_2 required to produce 200 cm³ of SO_3 = 100cm³.

Volume of air required = 100/20 x 100 = **500 cm³**

Exercise 6.6

a) *Calculate the mass of magnesium required to generate 200 cm³ H_2 at STP?*

$$Mg_{(s)} + 2HCl_{(aq)} \rightarrow MgCl_{2\,(aq)} + H_{2\,(g)}$$

Answer:

	$Mg_{(s)}$	+	$2HCl_{(aq)}$	→	$MgCl_{2(aq)}$	+	$H_{2(g)}$
A_r/ M_r	24		36.5		95		2
Reaction Coefficients	1		2		1		1
Volume (cm³)							200
Mass (g)	8.9 x 10⁻³ x 24 = **0.214**						
No. of moles	8.9 x 10⁻³						8.9 x 10⁻³

Number of moles in 200 cm³ of H_2 = volume / V_m = 200/22400 = 8.9 x 10⁻³.

According to the reaction coefficients in the balanced chemical equation, 1 mole of H_2 is generated from 1 mole of Mg.

Therefore, 200 cm³ of H_2 will be generated from 8.9 x 10^{-3} moles of Mg.

Mass of 8.9 x 10^{-3} moles of Mg = mass x M_r = 8.9 x 10^{-3} x 24 = **0.214 g**

b) Determine the volume of ozone, at STP, generated by reacting 0.95 g of fluorine with excess steam.

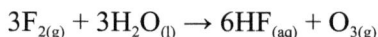

$$3F_{2(g)} + 3H_2O_{(l)} \rightarrow 6HF_{(aq)} + O_{3(g)}$$

Answer:

	$3F_2(g)$	+	$3H_2O_{(l)}$	→	$6HF_{(aq)}$	+	$O_{3(g)}$
Ar/ Mr	38		18		20		48
Reaction Coefficients	3		2		1		1
Volume (L)							185.9
Mass (g)	0.95						
No. of moles	0.96/38= 0.025						0.025/3 = 8.3 x 10^{-3}

Number of moles in 0.95 g of F_2 = mass/M_r = 0.95/38= 0.025.

According to the reaction coefficients in the balanced chemical equation, 3 moles of F_2 generate 1 mole of O_3.

Therefore, 0.025 moles F_2 will generate 0.025/3 = 8.3 x 10^{-3} moles of O_3.

Volume occupied by 8.3 x 10^{-3} moles of O_3 = number of moles x V_m = 8.3 x 10^{-3} x 22400 = **185.9 cm³**

c) Lead sulphide is converted to lead (II) oxide when heated in oxygen.

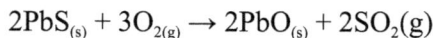

$$2PbS_{(s)} + 3O_{2(g)} \rightarrow 2PbO_{(s)} + 2SO_2(g)$$

Determine the volume of oxygen (molar volume = 24L) that would react exactly with 95.72 g of PbS?

Answer:

	$2PbS_{(s)}$	+	$3O_{2(g)}$	→	$PbO_{(s)}$	+	$SO_{22(g)}$
A_r/ M_r	239		32		223		64
Reaction Coefficients	2		3		1		1
Volume (cm³)			14,400				
Mass (g)	95.72						
No. of moles	0.4		0.6		0.2		0.2

Number of moles in 95.72 g of PbS = 95.72/239 = 0.4.

According to the reaction coefficients in the balanced chemical equation, 1 mole of PbS reacts with 1.5 (3/2) moles of O_2 to form 1 mole of PbO and 1 mole of O_2.

Therefore, 0.4 moles of PbS requires 1.5 x 0.4 = 0.6 moles of O_2.

Volume occupied by 0.6 moles of O_2 = number of moles x V_m = 0.6 x 24 = 14.4 dm³ = **14,400 cm³**

d) The chemical weapon phosgene is produced industrially according to the following reaction:

$$CO_{(g)} + Cl_{2(g)} → COCl_{2(g)}$$

Calculate the volume of CO (at STP) required to produce 10 g of phosgene?

Answer:

	$CO_{(g)}$	+	$Cl_{2(g)}$	→	$COCl_{2(g)}$
A_r/ M_r	28		71		99
Reaction Coefficients	1		1		1
Volume (cm³)	2,262				
Mass (g)					10
No. of moles	0.101				0.101

Number of moles in 10 g of $COCl_2$ = mass/ M_r = 10/99 = 0.101.

According to the reaction coefficients in the balanced chemical equation, 1 mole of $COCl_2$ is produced from 1 mole of CO.

Therefore, the number of moles of CO required to generate 0.101 moles of $COCl_2$ = 0.101

Volume occupied by 0.101 moles of CO = number of moles x V_m = 0.101 x 22,400 = **2,262 cm³**

e) What mass of ammonium chlorate is needed to decompose to give off 100 litres of oxygen at STP?

$$2KClO_{3\,(s)} \rightarrow 2KCl_{(s)} + 3O_2\,(g)$$

Answer:

	2KClO₃ (s)	→	**2KCl (s)**	+	**3O₂(g)**
A,/M,	122.5		74.5		32
Reaction Coefficients	2		2		2
Mass (g)	122.5 x 2.98 = **364.6g**				
Volume (L)					100
No. of moles	4.46 x 2/3 = 2.98		2.98		100/22.4 = 4.46

Number of moles of O_2 in 100 L = volume / V_m = 100/22.4 = 4.46.

According to the reaction coefficients in the balanced chemical equation, 2 moles of $KClO_3$ decompose to produce 3 moles of O_2.

Therefore, the number of moles O_2 produced from decomposition of 4.46 moles of $KClO_3$ = 4.46 x 2/3 = 2.98.

Mass of 2.98 moles of $KClO_3$ = number of moles x M_r = 2.98 x 122.5 = **364.6 g**

f) What volume of oxygen is required to completely combust a kilogram of octane at RTP?

Answer:

	2C₈H₁₈(l)	+	**25O₂(aq)**	→	**16CO₂(g)**	+	**18H₂O(l)**
A, /M,	114		32		44		18
Reaction coefficients	2		25		4		6
Volume (L)			109.65 x 24 = **2631.6**				

Table contd.....

	$2C_8H_{18(l)}$	+	$25O_{2(aq)}$	→	$16CO_{2(g)}$	+	$18H_2O_{(l)}$
Mass (g)	1,000						
No. of Moles	1000/114 = 8.77		25/2 x 8.77 = 109.65				

At RTP, $V_m = 24$ L

Number of moles in 1 kg (1000 g) of C_8H_{18} = mass/M_r = 1000/114 = 8.77

According to the reaction coefficients in the balanced chemical equation, 2 moles of C_8H_{18} require 25 moles of O_2 for total combustion.

Therefore, 8.77 moles of C_8H_{18} requires 25/2 x 8.77 = 109.65 moles of O_2 for total combustion.

Volume occupied by 109.65 moles of O_2 at STP = number of moles x V_m
= 109.64 x 24 = **2361.6 L**

g) How many grams of sodium do you have to put into water to make 30 litres of hydrogen at STP?

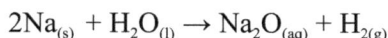

$$2Na_{(s)} + H_2O_{(l)} \rightarrow Na_2O_{(aq)} + H_{2(g)}$$

Answer:

	$2Na_{(s)}$	+	$H_2O_{(l)}$	→	$Na_2O_{(aq)}$	+	$H_{2(g)}$
A_r/M_r	23		18		62		18
Reaction Coefficients	2		1		1		1
Volume (L)							30
Mass (g)	2.68 x 23 = 61.6						
No. of Moles	2 x 1.34 = 2.68						30/22.4 = 1.34

Number of moles in 30 L of H_2 = 30/22.4 = 1.34.

According to the reaction coefficients in the balanced chemical equation, 2 moles of Na produce 1 mole of H_2.

Therefore number of moles of Na required to generate 1.34 moles of H_2 = 2 x 1.34 = 2.68.

Mass of 2.68 moles of Na = number of moles x A_r = 2.68 x 23 = **61.6 g**

h) Calculate the volume of oxygen produced at STP from the decomposition of

1.7g of hydrogen peroxide:

$$2H_2O_{2(aq)} \rightarrow 2H_2O_{(l)} + O_{2(g)}$$

Answer:

	$2H_2O_{2}$	$\rightarrow$	$2H_2O_{(l)}$	+	$O_{2(g)}$
A_r/M_r	34		18		32
Reaction Coefficients	2		2		2
Mass (g)	1.7				0.8
Volume (L)					0.025 x 22.4 = **0.56**
No. of moles	1.7/34 = 0.05				0.05/2 = 0.025

Number of moles in 1.7 g of H_2O_2 = mass/M_r = 1.7/34 = 0.05.
According to the reaction coefficients in the balanced chemical equation,
2 moles of H_2O_2 decompose to produce 1 mole of O_2.
Therefore, 0.05 moles of H_2O_2 will produce 0.5 x 0.05 = 0.025 moles of O_2.
Volume occupied by 0.025 moles of O_2 = number of moles x V_m = 0.05 x 22.4 = **0.56 L**

i) Sulphur dioxide can be removed from the waste gases of a power station by passing it through a slurry of calcium hydroxide. The equation for this reaction is:

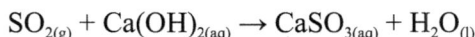

$$SO_{2(g)} + Ca(OH)_{2(aq)} \rightarrow CaSO_{3(aq)} + H_2O_{(l)}$$

What mass of calcium hydroxide would be needed to deal with 1000 dm³ of sulphur dioxide at STP?

Answer:

	$SO_{2(g)}$	+	$Ca(OH)_{2(aq)}$	$\rightarrow$	$CaSO_{3(aq)}$	+	$H_2O_{(l)}$
M_r	64		74		120		18
Reaction coefficients	1		1		1		1
Volume (dm³)	1000						
Mass (g)			4.64 x 74 = **3303.6**				

Table contd.....

	SO$_{2(g)}$	**+**	**Ca(OH)$_{2(aq)}$**	**→**	**CaSO$_{3(aq)}$**	**+**	**H$_2$O$_{(l)}$**
No. of Moles	1000/22.44 = 4.64		4.64				

Number of moles in 1000 dm^3 SO$_2$ = Volume/ V$_m$ = 1000/22.4 = 44.64.

According to the reaction coefficients in the balanced chemical equation, 1 mole of SO$_2$ is removed by 1 mole of Ca(OH)$_2$

Thus 44.64 moles of Ca(OH)$_2$ can remove 44.64 moles of SO$_2$.

Mass of 44.64 moles of Ca(OH)$_2$ = number of moles x M$_r$ = 44.64 x 74 = **3303.6 g**

j) What volume of oxygen, at STP, is required to completely combust 90 g ethane?

Answer:

	2C$_2$H$_{6(g)}$	**+**	**7O$_{2(g)}$**	**→**	**4CO$_{2(g)}$**	**+**	**6H$_2$O$_{(l)}$**
A$_r$/M$_r$	30		32		44		18
Reaction coefficients	2		7		4		6
Volume (L)			22.4 x 10.5 = **235.2**				
Mass (g)	90		336				
No. of Moles	90/30 = 3		7/2 x 3 = 10.5				

M$_r$ [C$_2$H$_6$] = 30.

Number of moles in 90 g of ethane = mass/M$_r$ = 90/30 = 3.

According to the reaction coefficients in the balanced chemical equation, 2 moles of C$_2$H$_6$ are totally combusted by 7 moles of O$_2$.

Therefore, 3 moles of C$_2$H$_6$ requires 3 x 7/2 = 10.5 moles of O$_2$ for total combustion.

Volume occupied by 10.5 moles O$_2$ = number of moles x V$_m$ = 10.5 x 22.4 = **235.2 L**

Mole Driving Test No. I

Counting Moles ....

M Driving Test M

To Pass the Mole Driving Test and throw away your **M**-plates you will need to achieve a score of greater than 70%. Each question is equally weighted (5%).

Relative Atomic Masses:
Hydrogen (H) = 1, Carbon (C) = 12, Nitrogen (N) = 14, Oxygen (O) = 16, Sodium (Na) = 23, Sulfur (S) = 32, Chlorine (Cl) = 35.5, Potassium (39), Calcium (Ca) = 40, Copper (Cu) = 63.5, Barium (Ba) = 137, Mercury(Hg) 200.5, Lead (Pb) = 207.

Molar Volume: 22.4 L or $22,4000 \text{ cm}^3$ at STP (273.15 K and 1 atm)

Avogadro's Number: 6.02×10^{23}

1. Rubidium has two naturally occurring two isotopes. 72.2% of all rubidium atoms have a relative atomic mass of 85 and the rest have a relative atomic mass of 87. What is the average relative atomic mass of rubidium?

2. Compound A was found to have the following composition: 47.0 % potassium, 14.5 % carbon, and 38.5 % oxygen and a relative molecular molar mass of 166.22. Determine the empirical and the molecular formula of compound A.

3. Lead (IV) oxide reacts with concentrated hydrochloric acid as follows:

$$PbO_{2(s)} + 4HCl_{(aq)} \rightarrow PbCl_{2(s)} + Cl_{2(g)} + 2H_2O_{(l)}$$

Calculate the mass of both lead chloride and chlorine gas that could be produced from 37.2g of PbO_2.

4. In 1774 Joseph Priestly produced oxygen by heating a sample of mercury II oxide with a large lens, according to the following equation:

$$2HgO(s) \rightarrow 2Hg(l) + O_{2(g)}$$

Calculate the volume of O_2 that could be generated from 1.08g of mercury (II) oxide. Assume that 1 mole of a gas occupies 24 dm³ under the experimental conditions.

Nigel P. Freestone
All rights reserved-© 2016 BENTHAM Science Publishers

5. Copper (II) nitrate thermally decomposes:

$$2Cu(NO_3)_{2(s)} \rightarrow 2CuO_{(s)} + 4NO_{2(g)} + O_{2(g)}$$

 a. What mass of copper (II) oxide could be produced from the thermal decomposition of 20.0 g of copper (II) nitrate?

 b. Calculate the mass of NO_2 produced?

6. Calculate the concentration of a solution containing 0.732 moles of ammonia in 250 cm^3 of solution.

7. Sodium hydride reacts violently with water forming flammable hydrogen gas:

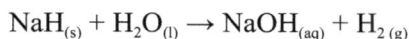

$$NaH_{(s)} + H_2O_{(l)} \rightarrow NaOH_{(aq)} + H_{2\,(g)}$$

A 1.00 g sample of sodium hydride was added to water and the resulting solution was diluted to a volume of exactly 250 cm^3

 a. What is the concentration sodium hydroxide solution formed?

 b. What volume of hydrogen gas, measured at STP, was generated?

 c. What volume of 0.112 M hydrochloric acid would be required to react exactly with a 25.0 cm^3 sample of the sodium hydroxide solution?

8. The hydrates of sodium carbonate can be represented by the general formula $Na_2CO_3.xH_2O$. A student dissolved a 3.01 g sample of one of these hydrates in water and made the solution up to 250 cm^3. In a titration, a 25.0 cm^3 aliquot of this solution required 24.3 cm^3 of 0.200 M hydrochloric acid for complete reaction.

The equation for this reaction is shown below.

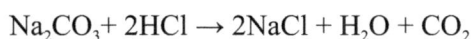

$$Na_2CO_3 + 2HCl \rightarrow 2NaCl + H_2O + CO_2$$

 a. Calculate the number of moles of HCl in 24.3 cm^3 of 0.200 M hydrochloric acid.

 b. Determine the number of moles of Na_2CO_3 present in 25.0 cm^3 of the Na_2CO_3 solution.

c. Deduce the number of moles of Na_2CO_3 in the original 250 cm^3 of solution.

d. Calculate the relative formula mass (M_r) of the hydrated sodium carbonate and hence the value of x.

9. A student prepared a calcium hydroxide solution by adding 0.00131 mole of calcium to a beaker containing about 100 cm^3, according to the reaction:

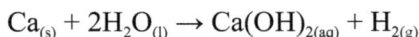

$$Ca_{(s)} + 2H_2O_{(l)} \rightarrow Ca(OH)_{2(aq)} + H_{2(g)}$$

a. Calculate the mass of calcium that the student added to the beaker of water.

b. What volume of hydrogen gas, at STP, would be generated?

c. The contents of the beaker were transferred to a 250 cm^3 volumetric flask and water was added to make the solution up to 250 cm^3. What is the concentration of hydroxide ions in the volumetric flask?

10. What is the volume of oxygen gas, at STP, required for the complete combustion of 1.0 kg of methane.

11. A 5.175 g sample of lead when heated in air at 300°C produced 5.708 g of an oxide of lead as the only product. Write a balanced chemical equation for this reaction.

12. 4.90 g of pure sulfuric acid was dissolved in water and the resulting solution was made up to 200 cm^3. On titration, a 10.0 cm^3 aliquot of a sodium hydroxide solution was completely neutralized by 20.7 cm^3 of this solution.

What is the concentration of the sodium hydroxide solution?

Mole Driving Test No. II

M Counting Moles ....
Driving Test **M**

To Pass the Mole Driving Test and throw away your **M**-plates you will need to achieve a score of greater than 70%. Each question is equally weighted (5%).

Relative Atomic Masses:
Hydrogen (H) = 1, Carbon (C) = 12, Nitrogen (N) = 14, Oxygen (O) = 16, Sodium (Na) = 23,
Magnesium (Mg = 24, Silicon (Si) = 28, Sulfur (S) = 32, Chlorine (Cl) = 35.5, Lead (Pb) = 207

Molar Volume: 22.4 L (22,400 cm³) at STP (273.15 K and 1 atm)

Avogadro's Number: 6.02×10^{23}

1. Calculate the average relative atomic mass of titanium given that it has five common isotopes, ^{46}Ti (8.0%), ^{47}Ti (7.8%), ^{48}Ti (73.4%), ^{49}Ti (5.5%), ^{50}Ti (5.3%)

2. Calculate the empirical and molecular formula of a compound of containing 17.04% Na, 47.41% S, with a relative formula mass 270.

3. In the sixteenth century, a large deposit of graphite was discovered in the Lake District. People at the time thought that the graphite was a form of lead. Graphite used in pencils is still referred to as 'pencil lead'. A student found that the mass of the 'pencil lead' in a school pencil was 0.321 g.

   a. How many moles of carbon atoms are present in 0.321g of student's pencil lead? Assume that the 'pencil lead' is pure graphite.

   b. Determine the number carbon atoms are present in the student's 'pencil lead'.

4. 19.6 g of hydrogen chloride, HCl, was dissolved in water and the volume made up to 250 cm³. What is the concentration of the resulting solution?

5. Determine the mass of H_2O required to completely react with 5.0 g of $SiCl_4$:

Nigel P. Freestone
All rights reserved-© 2016 Bentham Science Publishers

$$SiCl_4 + 2H_2O \rightarrow SiO_2 + 4HCl$$

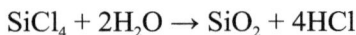

6. Magnesium carbonate and hydrochloric acid react according to the following equation:

$$MgCO_3 + 2HCl \rightarrow MgCl_2 + H_2O + CO_2$$

When a student added 75.0 cm^3 of 0.500 M hydrochloric acid to 1.25 g of impure $MgCO_3$ some acid was left unreacted. 21.6 cm^3 of a 0.500 M solution of sodium hydroxide was required to neutralize the unreacted acid.

   a. How many moles of HCl are present in 75.0 cm^3 of 0.500 M hydrochloric acid?

   b. How many moles of NaOH were used to neutralize the unreacted HCl.

   c. Determine the number of moles of HCl that reacted with the $MgCO_3$ in the sample.

   d. Calculate both the number of moles and the mass of $MgCO_3$ in the sample, and hence its purity.

7. The explosive nitroglycerine, $C_3H_5N_3O_9$, decomposes rapidly on detonation to form a large volume of gas, according to the following equation:

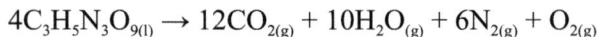

$$4C_3H_5N_3O_{9(l)} \rightarrow 12CO_{2(g)} + 10H_2O_{(g)} + 6N_{2(g)} + O_{2(g)}$$

0.350 g of oxygen gas was produced from the detonation of a sample of nitroglycerine.

   a. Determine the number of moles of oxygen gas produced from the detonation of nitroglycerine, and the total number of moles of gas generated.

   b. Determine the number of moles, and the mass, of nitroglycerine detonated.

8. Lead(II) sulfate can be produced from the reaction between lead nitrate and dilute sulfuric acid. What is the maximum amount of lead sulfate that could be obtained from 10 g of lead nitrate dissolved in water?

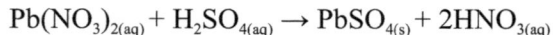

$$Pb(NO_3)_{2(aq)} + H_2SO_{4(aq)} \rightarrow PbSO_{4(s)} + 2HNO_{3(aq)}$$

9. The reaction between ammonium sulfate and aqueous sodium hydroxide is shown by the equation below.

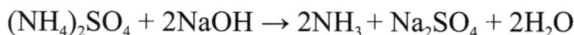

$$(NH_4)_2SO_4 + 2NaOH \rightarrow 2NH_3 + Na_2SO_4 + 2H_2O$$

A sample of ammonium sulfate was heated with 100 cm³ of 0.500 M aqueous sodium hydroxide. An excess of sodium hydroxide was used to ensure that all of the ammonium sulfate reacted. The unreacted sodium hydroxide required 27.3 cm³ of 0.600 M hydrochloric acid for neutralisation.

a. Calculate the original number of moles of NaOH in 100 cm³ of 0.500 M aqueous sodium hydroxide.

b. How many moles of HCl are present in 27.3 cm³ of 0.600 M hydrochloric acid?

c. Deduce the number of moles of the unreacted NaOH neutralized by the hydrochloric acid.

d. How many moles of NaOH reacted with the ammonium sulfate?

e. Determine the number of moles and the mass of ammonium sulfate in the sample.

10. The purity of commercially available sodium hydrogencarbonate was tested as follows. A 0.400g sample was dissolved in 100.0 cm³ of water and titrated against 0.200 M hydrochloric acid using methyl orange indicator.

$$NaHCO_3 + HCl \rightarrow NaCl + CO_2 + H_2$$

23.75 cm³ of acid was required for complete neutralisation.

a. How many moles of acid were used in the titration?

b. Calculate the mass of sodium hydrogen carbonate titrated and hence the purity of the sample.

Mole Driving Test No. III

Counting Moles ....
M Driving Test M

To Pass the Mole Driving Test and Throw Away Your **M**-plates You Will Need to Achieve a Score of Greater than 70%. Each Question is Equally Weighted (5%).

Relative Atomic Masses:
Hydrogen (H) = 1, Carbon (C) = 12, Nitrogen (N) = 14, Oxygen (O) = 16, Sodium (Na) = 23, Magnesium (Mg) = 24, Silicon (Si) = 28, Sulfur (S) = 32, Chlorine (Cl) = 35.5, Potassium (K) = 39, Titanium (Ti) = 48, Barium (Ba) = 137

Molar Volume: 22.4 L (22,4000 cm^3) at STP (273.15 K and 1 atm)

Avogadro's Number: 6.02 x 1023 mol^{-1}

1. Calculate the mass, in grams, of a single atom of sodium-23.

2. An organic compound was found to contain 12.8% carbon, 2.13% hydrogen and 85.07% bromine. If the compound has a relative molecular mass of 188, determine its empirical formula and the molecular formula.

3. What the mass of phosphorus is required to produce 200 g of phosphine (PH_3)?

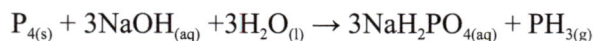

$$P_{4(s)} + 3NaOH_{(aq)} + 3H_2O_{(l)} \rightarrow 3NaH_2PO_{4(aq)} + PH_{3(g)}$$

4. A student heats 5.29g of $Sr(NO_3)_2$ and collects the gas.

$$2Sr(NO_3)_{2(s)} \rightarrow 2SrO_{(s)} + 4NO_{2(g)} + O_{2(g)}$$

Determine the volume of gas, at STP, obtained by the student. Molar mass of $Sr(NO_3)_2$ = 211.6 g mol^{-1}.

5. A 12.41 g sample of hydrated sodium thiosulfate, $Na_2S_2O_3 \cdot 5H_2O$, was heated to remove the water of crystallization.

    a. What is the relative formula mass of $Na_2S_2O_3 \cdot 5H_2O$?

Nigel P. Freestone
All rights reserved-© 2016 Bentham Science Publishers

    b. Calculate the expected mass of anhydrous sodium thiosulfate that forms.

6. The Kroll process is used to convert ore into titanium metal. Titanium chloride produced from the chlorination of ore, is reduced to titanium metal using magnesium under an inert atmosphere.

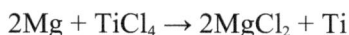

$$2Mg + TiCl_4 \rightarrow 2MgCl_2 + Ti$$

Calculate the maximum mass of titanium that could be produced from the addition of 3800 kg of titanium chloride to 1500 kg of magnesium.

7. A 1 mg sample of octane, C_8H_{18} was totally combusted in air.

    a. How many moles are present in 1 mg of octane ?

    b. Determine the number of moles and the volume of carbon dioxide generated.

8. Metal **M** forms a carbonate (M_2CO_3), which reacts with hydrochloric acid according to the following equation:

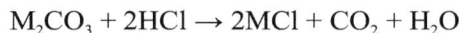

$$M_2CO_3 + 2HCl \rightarrow 2MCl + CO_2 + H_2O$$

A 0.394 g sample of M_2CO_3, was found to require the addition of 21.7 cm^3 of a 0.263 M solution of hydrochloric acid (HCl) for complete reaction.

    a. Determine the number of moles of hydrochloric acid added.

    b. Determine the relative molecular mass of M_2CO_3 and hence the identity of M

9. An ammonia solution was reacted with sulfuric acid:

$$2NH_{3(aq)} + H_2SO_{4(aq)} \rightarrow (NH_4)_2SO_{4(aq)}$$

A 25.0 cm^3 aliquot of 1.24 M sulfuric acid required 30.8 cm^3 of this ammonia solution for complete reaction.

Calculate the concentration of the ammonia and the mass of ammonium sulfate present in the solution at the end of this titration.

10. Hydrogen is produced by the addition of hydrochloric acid to magnesium metal:

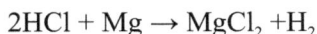

$$2HCl + Mg \rightarrow MgCl_2 + H_2$$

What mass of hydrogen is produced from the addition of 100 cm^3 of 5M hydrochloric acid of to an excess of magnesium ?

11. Ammonium nitrate is produced industrially by the reaction between ammonia and nitric acid:

$$NH_3 + HNO_3 \rightarrow NH_4NO_3$$

Calculate the volume of 2M nitric acid required to react with exactly 20.0 g of ammonia.

12. Potassium chlorate, $KClO_3$, thermally decomposes according to the following equation:

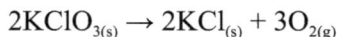

$$2KClO_{3(s)} \rightarrow 2KCl_{(s)} + 3O_{2(g)}$$

a. What mass of oxygen could be produced from the complete decomposition of 1.47 g of $KClO_3$?

b. What mass of $KClO_3$ is required to generate 1.00 dm^3 of oxygen at STP ?

13. 25.0 cm^3 of 0.25 M sodium hydroxide required 22.5 cm^3 of a hydrochloric acid solution for complete neutralisation. Calculate the concentration of the HCl solution.

14. Barium nitrate thermally decomposes as follows:

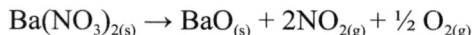

$$Ba(NO_3)_{2(s)} \rightarrow BaO_{(s)} + 2NO_{2(g)} + \tfrac{1}{2} O_{2(g)}$$

a. What is the total volume of gas, at STP, generated by the thermal decomposition of 5.00 g of barium nitrate ?

b. What volume of 1.20 M hydrochloric acid is required to neutralize the barium oxide produced by the thermal decomposition of 5.00 g of barium nitrate. Barium oxide reacts with hydrochloric acid as follows:

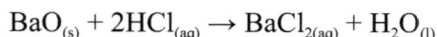

$$BaO_{(s)} + 2HCl_{(aq)} \rightarrow BaCl_{2(aq)} + H_2O_{(l)}$$

15. Silver nitrate thermally decomposes. 0.720 g of silver and 0.307 g of nitrogen dioxide was produced from heating a 1.133 g sample of silver nitrate in an open tube. The rest of the mass loss was due to oxygen. Using this data write the balanced chemical equation for the reaction.

Mole Driving Test Answers

M Counting Moles .... **Driving Test** **M**

To Pass the Mole Driving Test and Throw Away Your **M**-plates You Will Need To Achieve a Score of Greater than 70%. Each Question is Equally Weighted (5%).

Relative Atomic Masses:

Hydrogen (H) = 1, Carbon (C) = 12, Nitrogen (N) = 14, Oxygen (O) = 16, Sodium (Na) = 23, Sulfur (S) = 32, Chlorine (Cl) = 35.5, Potassium (39), Calcium (Ca) = 40, Copper (Cu) = 63.5, Barium (Ba) = 137, Mercury(Hg) 200.5, Lead (Pb) = 207.

Molar Volume: 22.4 L or 22,4000 cm³ at STP (273.15 K and 1 atm)

Avogadro's Number: 6.02 x 10²³

1.   Rubidium has two naturally occurring two isotopes. 72.2% of all rubidium atoms have a relative atomic mass of 85 and the rest have a relative atomic mass of 87. What is the average relative atomic mass of rubidium?

Answer:

Average Atomic Mass=(72.2/100 x 85) + (27.8/100 x 87)
=61.37 + 24.186
=**85.56**

2.   Compound A was found to have the following composition: 47.0 % potassium, 14.5 % carbon, and 38.5 % oxygen and a relative molecular molar mass of 166.22. Determine the empirical and the molecular formula of compound A.

Answer:

	Potassium	Carbon	Oxygen
% composition	47.0	14.5	38.5
A$_r$	39	12	16
% composition/A$_r$	47/39 = 1.2	14.5/12 = 1.2	38.5/16 = 2.4

Nigel P. Freestone
All rights reserved-© 2016 Bentham Science Publishers

Table contd.....

	Potassium	Carbon	Oxygen
Ratio	1	1	2

Empirical formula: KCO_2

Relative formula mass $KCO_2 = (1 \times 39) + 12 + (2 \times 16) = 83$

Molecular formula $= (KCO_2)_n$

$n = 166/83 = 2$

Molecular formula: $(KCO_2)_2 = \textbf{K}_2\textbf{C}_2\textbf{O}_4$

3. Lead (IV) oxide reacts with concentrated hydrochloric acid as follows:

$$PbO_{2(s)} + 4HCl_{(aq)} \rightarrow PbCl_{2(s)} + Cl_{2(g)} + 2H_2O_{(l)}$$

Calculate the mass of both lead chloride and chlorine gas that could be produced from 37.2 g of PbO_2.

Answer:

	$PbO_{2(s)}$	+	$4HCl_{(aq)}$	→	$PbCl_{2(s)}$	+	$Cl_{2(g)}$	+	$2H_2O_{(l)}$
M_r	239		36.5		278		71		18
Mass Balance	239		146		278		71		36
		385				385			
Reaction Coefficients	1		4		1		1		1
Mass (g)	37.2				0.1556 x 278 = **43.27**		0.1556 x 71 = **11.05**		
No. of Moles	0.1556				0.1556		0.1556		

Number of moles in 37.2 g of PbO_2 = mass/M_r = 347.2/239 = 0.1556

According to the balanced chemical equation, 1 mole of PbO_2 produces 1 mole $PbCl_2$

Therefore, 0.1556 moles of PbO_2 can produce 0.1556 moles of $PbCl_2$

Mass of 0.1556 moles of $PbCl_2$ = number of moles x M_r = 0.1556 x 278 = 43.27 g

Mass of $PbCl_2$ = **43.27 g**

According to the balanced chemical equation, 1 mole of PbO_2 produces 1 mole of Cl_2

Therefore, 0.1556 moles of PbO_2 can produce 0.1556 moles of Cl_2

Mass of 0.01556 moles of Cl_2 = number of moles x M_r = 0.1556 x 71 = 11.05 g

Mass of Cl_2 = **11.05 g**

4.   In 1774 Joseph Priestly produced oxygen by heating a sample of mercury II oxide with a large lens, according to the following equation:

$$2HgO(s) \rightarrow 2Hg(l) + O_{2(g)}$$

Calculate the volume of O_2 that could be generated from 1.08g of mercury (II) oxide. Assume that 1 mole of a gas occupies 24 dm^3 under the experimental conditions.

Answer:

	2HgO$_{(s)}$	**→**	**2Hg$_{(l)}$**	**+**	**O$_{2(g)}$**
A$_r$/M$_r$	216.5		200.5		32
Mass Balance	433		401		32
				433	
Reaction Coefficients	2		2		1
Mass	1.08				
No. of Moles	1.08/216.5 = 0.005		0.005		0.0025
Volume (L)					0.025 x 22.4 = **0.06**

Number of moles in 1.08 g of HgO = 1.08/216.5 = 0.005
According to the balanced chemical equation 2 moles of HgO generates 1 mole of O_2
Thus number of moles of O_2 produced from 0.005 moles of HgO = 0.005/2 = 0.0025
Volume of 0.0025 moles of O_2 = 0.0025 x 24 = **0.06 L = 60 cm^3**

5.   Copper (II) nitrate thermally decomposes:

$$2Cu(NO_3)_{2(s)} \rightarrow 2CuO_{(s)} + 4NO_{2(g)} + O_{2(g)}$$

a)   What mass of copper (II) oxide could be produced from the thermal decomposition of 20.0 g of copper (II) nitrate?

Answer:

	2Cu(NO$_3$)$_{2(s)}$	**→**	**2CuO(s)**	**+**	**4NO$_{(g)}$**	**+**	**O$_{2(g)}$**
M$_r$	187.5		79.5		46		32
Mass Balance			159		184		32
	375			375			

Table contd.....

	$2Cu(NO_3)_{2(s)}$	→	$2CuO(s)$	+	$4NO_{(g)}$	+	$O_{2(g)}$
Reaction Coefficients	2		2		4		1
Mass (g)	20		0.1066 x 79.5 = **8.48**		0.2013 x 46 = **9.8**		
No. of Moles	20/187.5 = 0.1066		0.1066		2 x 0.1066 = 0.2013		

Number of moles in 20 g of $Cu(NO_3)_2$ = mass/M_r = 20/187.5 = 0.1066

According to the balanced chemical equation, 1 mole of $Cu(NO_3)_2$ produces 1 mole of CuO

Therefore, 0.1066 moles of $Cu(NO_3)_2$ could produce 0.1066 moles of CuO

Mass of 0.1066 moles of CuO = number of moles x M_r = 0.1066 x 79.5 = 8.48 g

Mass of CuO = **8.48 g**

b)   Calculate the mass of NO_2 produced?

Answer:

According to the balanced chemical equation, 1 mole of $Cu(NO_3)_2$ produces 2 moles of NO_2

Therefore, 0.1066 moles of $Cu(NO_3)_2$ could produce 0.2132 moles (2 x 0.1066) of NO_2

Mass of 0.2132 moles of NO_2 = number of moles x M_r = 0.2132 x 46 = 9.8 g

Mass of NO_2 = **9.8 g**

6.   Calculate the concentration of a solution containing 0.732 moles of ammonia in 250 cm^3 of solution.

Answer:

Convert volume from cm^3 to litres

Volume = 250 cm^3 = 0.25 litres

Number of moles of NH_3 = 0.732

Concentration = number of moles/volume (L) = 0.732/0.25 = 2.928 M

Concentration of ammonia solution = **2.928 M**

7.   Sodium hydride reacts violently with water forming flammable hydrogen gas:

$$NaH_{(s)} + H_2O_{(l)} \rightarrow NaOH_{(aq)} + H_{2(g)}$$

A 1.00 g sample of sodium hydride was added to water and the resulting solution was diluted to a volume of exactly 250 cm^3.

a) What is the concentration sodium hydroxide solution formed?

Answer:

	NaH$_{(s)}$	+	H$_2$O$_{(l)}$	→	NaOH$_{(aq)}$	+	H$_{2(g)}$
M$_r$	24		18		40		2
Mass Balance		42				42	
Reaction Coefficients	1		1		1		1
Mass	1						
No. of Moles	1/24 = 0.0417						0.0417
Volume (L)	0.25						0.0417 x 22.4 = **0.934**
Concentration (M)							0.0417/0.25 = **0.167**

Number of moles in 1 g of NaH = 1/24 = 0.0417
According to the balanced chemical equation, 1 mole of NaH produces 1 mole of NaOH
Thus number of moles of NaOH produced from 0.0417 moles of NaH = 0.0417
Volume = 250 cm^3 = 0.25 L
Concentration = number of moles/volume = 0.0417/0.25
Concentration of NaOH = **0.167 M**

b) What volume of hydrogen gas, measured at STP, was generated?

Answer:

According to the balanced chemical equation, 1 mole of NaH produces I mole of H$_2$
Therefore, 0.01417 moles of NaH generates 0.01417 moles of H$_2$
Given that 1 mole of gas occupies a volume of 22.4 L at STP
Therefore, 0.01417 moles of H$_2$ will occupy a volume of 0.0417 x 22.4 = **0.934 L**

c) What volume of 0.112 M hydrochloric acid would be required to react exactly with a 25.0 cm^3 sample of the sodium hydroxide solution?

Answer:

	NaOH$_{(aq)}$	+	HCl$_{(aq)}$	→	NaCl + H$_2$O
Reaction Coefficients	1		1		
Volume (cm^3)	25		4.175 x 10^{-3} x 1000/0.112 = **37.28**		
No. of Moles	25/1000 x 0.167 = 4.175 x 10^{-3}		4.175 x 10^{-3}		
Concentration (M)	0.167		0.112		

Number of moles in 25 cm^3 of 0.167 M NaOH = 25/1000 x 0.167 = 4.175 x 10^{-3}

According to the balanced chemical equation, 1 mole of NaOH reacts with 1 mole of HCl

Therefore the number of moles HCl that reacted exactly with 4.175 x 10^{-3} moles of NaOH = 4.175 x 10^{-3}

Volume of 0.112M containing 4.175 x 10^{-3} moles of HCl = 4.175 x 10^{-3} x 1000/0.112 = **37.28 cm^3**

8.   The hydrates of sodium carbonate can be represented by the general formula Na$_2$CO$_3$.xH$_2$O. A student dissolved a 3.01 g sample of one of these hydrates in water and made the solution up to 250 cm^3. In a titration, a 25.0 cm^3 aliquot of this solution required 24.3 cm^3 of 0.200 M hydrochloric acid for complete reaction.

The equation for this reaction is shown below.

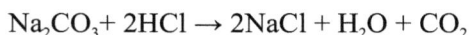

$$Na_2CO_3 + 2HCl \rightarrow 2NaCl + H_2O + CO_2$$

a)   Calculate the number of moles of HCl in 24.3 cm^3 of 0.200 M hydrochloric acid.

Answer:

	Na$_2$CO$_{3(aq)}$	+	2HCl$_{(aq)}$	→	2NaCl + CO$_2$ + H$_2$O
Reaction Coefficients	1		2		
Volume (cm^3)	25		24.3		
No. of Moles	4.86 x 10^{-3} /2 = **2.43 x 10^{-3}**		24.3/1000 x 0.2 = **4.86 x 10^{-3}**		
Concentration (M)	1000/25 x 2.43 x 10^{-3} = **0.0972**		0.2		

Number of moles in 24.3 cm^3 of 0.200 M HCl =volume (L) x concentration = 24.3/1000 x 0.200 = **4.86 x 10^{-3}**

b)   Determine the number of moles of Na_2CO_3 present in 25.0 cm^3 of the Na_2CO_3 solution.

Answer:

According to the balanced chemical equation, at the end point, the number of moles of Na_2CO_3 = 0.5 x number of moles of HCl
Therefore number of moles of Na_2CO_3 in 25.0 cm^3 of the Na_2CO_3 solution = 0.5 x 4.86 x 10^{-3} = **2.43 x 10^{-3}**

c)   Deduce the number of moles of Na_2CO_3 in the original 250 cm^3 of solution.

Answer:

If number of moles of Na_2CO_3 present in 25 cm^3, concentration of the solution (*i.e* the number of moles in 1000 cm^3) = 1000/25 x 2.43 x 10^{-3} = **0.0972 M**

d)   Calculate the relative formula mass (M_r) of the hydrated sodium carbonate and hence the value of x.

Answer:

Number of moles of $Na_2CO_3 xH_2O$ in 250 cm^3 = 250/1000 = 0.0972 = 0.0243
Mass of 0.0243 moles of $Na_2CO_3xH_2O$ = 3.01 g
M_r = mass/number of moles
M_r [$Na_2CO_3xH_2O$] = 3.01/0.0243 = 123.9
Since M_r [Na_2CO_3] =106
M_r [xH_2O] = M_r [$Na_2CO_3xH_2O$] - M_r [Na_2CO_3] = 123.9 – 106 = 17.9; therefore **x =1**
Formula of hydrate = **$Na_2CO_3.H_2O$**

9. A student prepared a calcium hydroxide solution by adding 0.00131 mole of calcium to a beaker containing about 100 cm^3, according to the reaction:

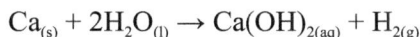

$$Ca_{(s)} + 2H_2O_{(l)} \rightarrow Ca(OH)_{2(aq)} + H_{2(g)}$$

a)   Calculate the mass of calcium that the student added to the beaker of water.

Answer:

	$Ca_{(s)}$	+	$2H_2O_{(l)}$	→	$Ca(OH)_{2(aq)}$	+	$H_{2(g)}$
M_r	40		18		74		2
Mass Balance	40		2 x 18 = 36		74		2
		76				76	
Reaction Coefficients	1		2		1		1
Mass (g)	0.00131 x 40 = **0.0524**						
No. of moles	0.00131						0.0131
Volume (cm³)							0.00131 x 22.4 = **0.293**

Mass of 0.00131 mole of calcium = number of moles x M_r = 0.00131 x 40 = **0.0524 g**

b)   What volume of hydrogen gas, at STP, would be generated?

Answer:

According to the balanced chemical equation, 1 mole of Ca produces 1 mole of H_2
Thus 0.00131 mole of Ca can generate 0.00131 mole of H_2
1 mole of gas at STP occupies a volume of 22.4 L (V_m)
Volume = Number of moles x V_m
Thus 0.00131 mole of H_2 will occupy a volume of 0.00131 x 22.4 = **0.293 L**

c)   The contents of the beaker were transferred to a 250 cm³ volumetric flask and water was added to make the solution up to 250 cm³. What is the concentration of hydroxide ions in the volumetric flask?

Answer:

According to the balanced chemical equation, 1 mole of Ca produces 1 mole of $Ca(OH)_2$
Thus 0.00131 mole of Ca can generate 0.00131 mole of $Ca(OH)_2$
Given that $Ca(OH)_2$ totally dissociates: $Ca(OH)_2 \rightarrow Ca^{2+} + 2OH^-$
The number of moles of OH^- ions in 250 cm³ = 2 x 0.00131 = 0.0262 moles
Concentration of OH^- ions = 1000/250 x 0.0262 = **0.0105 M**

10.   What is the volume of oxygen gas, at STP, required for the complete combustion of 1.0 kg of methane.

Answer:

	$CH_{4(g)}$	+	$2O_{2(g)}$	→	$CO_{2(g)}$	+	$2H_2O_{(l)}$
M_r	16		32		44		18
Mass Balance	16		64		44		36
		80				80	
Reaction Coefficients	1		2		1		1
Mass (g)	1000						
No. of moles	1000/16 = 62.5		62.5 x 2 = 125				
Volume (L)			125 x 22.4 = **2800**				

Number of moles of in 1000 g (1 kg) of CH_4 = 1000/16 = 62.5
According to the balanced chemical equation, 1 mole of CH_4 required 2 moles of O_2 for total combustion
Thus 62.5 moles of CH_4 requires 62.5 x 2 = 125 moles O_2 for total combustion
Volume occupied by 125 moles of O_2 at STP = 22.4 x 125 = **2,800 L**

11.   A 5.175 g sample of lead when heated in air at 300°C produced 5.708 g of an oxide of lead as the only product. Write a balanced chemical equation for this reaction.

Answer:

	$xPb_{(s)}$	+	$yO_{2(g)}$	→	$PbXO_{y(s)}$
M_r	207		32		
Reaction Coefficients	1		1.5		
Mass (g)	5.175		5.708-5.175 = 0.533		5.708
No. of Moles	0.025		0.0167		

Number of moles in 5.175 g of Pb = mass/M_r = 5.175/207 = 0.025 moles
Mass of oxygen consumed in the formation of the lead oxide = 5.708 – 5.175 g = 0.533 g
Number of moles of O_2 consumed = mass/M_r = 0.533/32 = 0.0167 moles
Thus 0.025 moles of Pb react with 0.0167 moles of O_2 to give product
Thus 1.5 moles (0.025/0.0167) of Pb react with 1 mole (0.0167/0.0167) of O_2 to produce the lead oxide
Therefore, 3 moles of Pb react with 2 moles of O_2 to give product, *i.e.*

3Pb + 2O$_2$ → Pb$_3$O$_4$

12. 4.90 g of pure sulfuric acid was dissolved in water and the resulting solution was made up to 200 cm^3. On titration, a 10.0 cm^3 aliquot of a sodium hydroxide solution was completely neutralized by 20.7 cm^3 of this solution.

What is the concentration of the sodium hydroxide solution?

Answer:

	H$_2$SO$_{4(aq)}$	+	2NaOH$_{(aq)}$	→	Na$_2$SO$_4$ + 2H$_2$O
Reaction Coefficients	1		2		
Volume (cm^3)	20.7		10		
No. of moles	20.7/1000 x 0.25 = 0.0052		2 x 0.0052 = 0.0104		
Concentration (M)	0.25		1000/10 x 0.0104 = **1.04**		

Number of moles in 4.9 g of H$_2$SO$_4$ = 4.9/98 = 0.05
Volume = 200 cm^3 = 0.2 L
Concentration = number of moles /volume = 0.05/0.2 = 0.25M
Number of moles in 20.4 cm^3 of 0.25M H$_2$SO$_4$ = volume x concentration = 20.4/1000 x 0.25 = 0.0052
At the end point, 2 x number of moles of H$_2$SO$_4$ = number of moles of NaOH in 10 cm^3
Thus, the number of moles of NaOH in 10cm^3 = 2 x 0.0052 = 0.0104
Concentration of NaOH =1000/10 x 0.0104 = **1.04M**

M   Counting Moles ....
Driving Test M

To Pass the Mole Driving Test and throw away your **M**-plates you will need to achieve a score of greater than 70%. Each question is equally weighted (5%).

Relative Atomic Masses:
Hydrogen (H) = 1, Carbon (C) = 12, Nitrogen (N) = 14, Oxygen (O) = 16, Sodium (Na) = 23, Magnesium (Mg = 24, Silicon (Si) = 28, Sulfur (S) = 32, Chlorine (Cl) = 35.5, Lead (Pb) = 207
Molar Volume: 22.4 L (22,400 cm³) at STP (273.15 K and 1 atm)
Avogadro's Number: 6.02 x 10^{23}

1. Calculate the average relative atomic mass of titanium given that it has five common isotopes, ^{46}Ti (8.0%), ^{47}Ti (7.8%), ^{48}Ti (73.4%), ^{49}Ti (5.5%), ^{50}Ti (5.3%)

Answer:

Average Atomic Mass = (8/100 x 46) + (7.8/100 x 47) + (73.4/100 x 48) + (5.5/100 x 49) + (5.3/100 x 50)

$$= 3.68 + 3.67 + 35.23 + 2.695 + 2.65$$
$$= 47.93$$

2. Calculate the empirical and molecular formula of a compound of containing 17.04% Na, 47.41% S, with a relative formula mass 270.

Answer:

	Sodium	Sulfur	Oxygen
% composition	17.04	47.41	35.25
A_r	23	32	16
% composition/A_r	17.04/23 = 0.74	47.41/32 = 1.48	35.25/16 = 2.2
Ratio	1	2	3

Empirical Formula: **NaS_2O_3**

$M_r[NaS_2O_3] = 135$

Molecular formula = $(NaS_2O_3)n$

$n = M_r$ [Molecular formula] $/M_r$ [Empiral formula] = 270/135 = 2

Molecular formula: $Na_2S_4O_6$

3. In the sixteenth century, a large deposit of graphite was discovered in the Lake District. People at the time thought that the graphite was a form of lead. Graphite used in pencils is still referred to as 'pencil lead'. A student found that the mass of the 'pencil lead' in a school pencil was 0.321 g.

a) How many moles of carbon atoms are present in 0.321g of student's pencil lead? Assume that the 'pencil lead' is pure graphite.

Answer:

A_r [C] = 12

Number of moles in 0.321 g of C = mass/A_r = 0.321/12 = **0.02675**

b) Determine the number carbon atoms are present in the student's 'pencil lead'.

Answer:

Number of particles (atoms, molecules *etc*) = number of moles x Avogadro's number

Number of carbon atoms in 0.02675 moles = 0.02675 x 6.02 x 10^{23} = **1.61 x 10^{22}**

4. 19.6 g of hydrogen chloride, HCl, was dissolved in water and the volume made up to 250 cm^3. What is the concentration of the resulting solution?

Answer:

M_r [HCl] = 36.5

Number of moles in 19.6 g of HCl = mass/M_r = 19.6/36.5 = 0.537

Number of moles of HCl in 250 cm^3 = 0.537

Convert volume to litres: 250 cm^3 = 0.25 L

Concentration of HCl = number of moles/volume = 0.536/0.25 = **2.15 M**

5. Determine the mass of H_2O required to completely react with 5.0 g of $SiCl_4$:

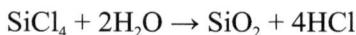

$$SiCl_4 + 2H_2O \rightarrow SiO_2 + 4HCl$$

Answer:

	SiCl$_4$	+	2H$_2$O	→	SiO$_2$	+	4HCl
M$_r$	170		18		60		36.5
Mass Balance	170	206	36		60	206	146
Reaction Coefficients	1		2		1		4
Mass (g)	5		0.0588 x 18 = **1.06**				
No. of Moles	5/170 = 0.0294		2 x 0.0294 = 0.0588				

Number of moles in 5 g of SiCl$_4$ = mass/M$_r$ = 5/170 = 0.0294
According to reaction coefficients, 1 mole of SiCl$_4$ reacts with 2 moles of H$_2$O
Thus 0.0294 moles of SiCl$_4$ will react with 2 x 0.0294 = 0.0588 moles of H$_2$O
Mass of H$_2$O = number of moles x M$_r$ = 0.0588 x 18 = **1.06 g**

6. Magnesium carbonate and hydrochloric acid react according to the following equation:

$$MgCO_3 + 2HCl \rightarrow MgCl_2 + H_2O + CO_2$$

When a student added 75.0 cm^3 of 0.500 M hydrochloric acid to 1.25 g of impure MgCO$_3$ some acid was left unreacted. 21.6 cm^3 of a 0.500 M solution of sodium hydroxide was required to neutralize the unreacted acid.

a) How many moles of HCl are present in 75.0 cm^3 of 0.500 M hydrochloric acid?

Answer:

Number of moles in 75.0 cm^3 of 0.500 M HCl = volume (L) x concentration = 75/1000 x 0.500 = **0.0375**

b) How many moles of NaOH were used to neutralize the unreacted HCl.

Answer:

Number of moles in 21.6 cm^3 of 0.500 M NaOH solution = volume (L) x concentration = 21.6/1000 x 0.5 = **0.0108 M**

c) Determine the number of moles of HCl that reacted with the MgCO$_3$ in the sample.

Answer:

Number of moles of HCl which reacted with the $MgCO_3$ = 0.0375 – 0.0108 = **0.0267**

d) Calculate both the number of moles and the mass of $MgCO_3$ in the sample, and hence its purity.

Answer:

M_r [$MgCO_3$] = 84
According to the balanced chemical equation, the number of moles of M_2CO_3 = 0.5 x number of moles of HCl
Thus the number of moles of $MgCO_3$ = 0.5 x 0.0267 = 0.01335
Mass of 0.01355 moles of $MgCO_3$ = number of moles x M_r = 0.01335 x 84 =1.12 g
% $MgCO_3$ in sample = 1.12/1.25 x 100 = **90%**

7. The explosive nitroglycerine, $C_3H_5N_3O_9$, decomposes rapidly on detonation to form a large volume of gas, according to the following equation:

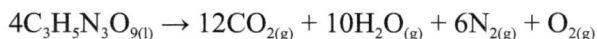

$$4C_3H_5N_3O_{9(l)} \rightarrow 12CO_{2(g)} + 10H_2O_{(g)} + 6N_{2(g)} + O_{2(g)}$$

0.350 g of oxygen gas was produced from the detonation of a sample of nitroglycerine.

a) Determine the number of moles of oxygen gas produced from the detonation of nitroglycerine, and the total number of moles of gas generated.

Answer:

	$4C_3H_5N_3O_{9(l)}$	$\rightarrow$	$12CO_{2(g)}$	+	$10H_2O_{(l)}$	+	$6N_{2(g)}$	+	$O_{2(g)}$
M_r	227		44		18		28		32
Mass Balance	908	908	528		180		168	908	32
Reaction Coefficients	4		12		10		6		1
Mass (g)	0.438 x 227 = 9.93								0.35
No. of Moles	4 x 1.09 x 10^{-3} = 0.0438								0.35/32 = 1.09 x 10^{-3}

Number of moles in 0.350 g of O_2 = mass/M_r = 0.350/32 = **0.0109**
One mole of nitroglycerine on detonation generates 29 (12 + 10 + 6 + 1) moles of gas
Total number of moles of gas produced = 29 x 0.0109 = **0.317**

b) Determine the number of moles, and the mass, of nitroglycerine detonated.

Answer:

According to reaction coefficients, 4 moles of nitroglycerine produce 1 mole of O_2
Therefore, 0.0109 moles of O_2 are generated from 4 x 0.0109 = 0.0438 mole of nitroglycerine
Mass of 0.0438 moles of nitroglycerine = number of moles x M_r = 0.0438 x 227 = **9.93 g**

8. Lead(II) sulfate can be produced from the reaction between lead nitrate and dilute sulfuric acid. What is the maximum amount of lead sulfate that could be obtained from 10 g of lead nitrate dissolved in water?

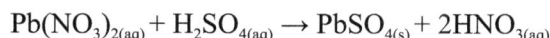

$$Pb(NO_3)_{2(aq)} + H_2SO_{4(aq)} \rightarrow PbSO_{4(s)} + 2HNO_{3(aq)}$$

Answer:

	$Pb(NO_3)_{2(aq)}$	+	$H_2SO_{4(aq)}$	$\rightarrow$	$PbSO_{4(s)}$	+	$2HNO_{3(aq)}$
M_r	331		98		303		63
Mass Balance		429			303	429	126
Reaction Coefficients	1		1		1		1
Mass (g)	10				**9.15**		
No. of Moles	0.0302				0.0302		

Number of moles in 10 g of $Pb(NO_3)_2$ = mass10/331 = 0.0302
According to the balanced chemical equation I mole of $Pb(NO_3)_2$ produces 1 mole of $PbSO_4$
Therefore, the number of moles of $PbSO_4$ generated from 10 g of $Pb(NO_3)_2$ = 0.0302 moles
Mass of 0.0302 moles of $PbSO_4$ = 0.0304 x 303 = **9.15 g**

9. The reaction between ammonium sulfate and aqueous sodium hydroxide is shown by the equation below.

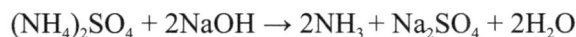

$$(NH_4)_2SO_4 + 2NaOH \rightarrow 2NH_3 + Na_2SO_4 + 2H_2O$$

A sample of ammonium sulfate was heated with 100 cm³ of 0.500 M aqueous sodium hydroxide. An excess of sodium hydroxide was used to ensure that all of the ammonium sulfate reacted. The unreacted sodium hydroxide required 27.3 cm^3 of 0.600 M hydrochloric acid for neutralisation.

Answer:

	$(NH_4)_2SO_4$	+	2NaOH	→	2NH$_3$	+	Na$_2$SO$_4$	+	2H$_2$O
M$_r$	132		40		17		142		18
Mass Balance	132	212	80		34		142	212	36
Reaction Coefficients	1		2		2		1		1
Mass	0.0168 x 132 = 2.22								
No. of moles	0.0336 x 0.5 = 0.0168		0.0366						
Volume (cm^3)									
Concentration (M)			0.5						

a) Calculate the original number of moles of NaOH in 100 cm^3 of 0.500 M aqueous sodium hydroxide.

Answer:

Number of moles of NaOH in 100 cm^3 of 0.500 M = volume (L) x concentration = 100/1000 x 0.500 = **0.05 moles**

b) How many moles of HCl are present in 27.3 cm^3 of 0.600 M hydrochloric acid?

Answer:

Number of moles in 27.3 cm^3 of 0.600 M HCl = volume (L) x concentration = 27.3/1000 x 0.600 = **0.0164**

c) Deduce the number of moles of the unreacted NaOH neutralized by the hydrochloric acid.

Answer:

According to the balanced chemical equation, 1 mole of HCl neutralizes 1 mole of NaOH
The number of mole of unreacted NaOH = **0.0164**

d) How many moles of NaOH reacted with the ammonium sulfate?

Answer:

Number of moles of NaOH reacted with ammonium sulfate = 0.05 − 0.0164 = **0.0336**

e) Determine the number of moles and the mass of ammonium sulfate in the sample.

Answer:

M_r [$(NH_4)_2SO_4$] = 132
According to the balanced chemical equation, 2 moles of NaOH react with 1 mole of $(NH_4)_2SO_4$
Therefore, 0.0336 moles of NaOH react with 0.5 x 0.336 = **0.0168** moles of $(NH_4)_2SO_4$
Mass of 0.0168 moles of $(NH_4)_2SO_4$ = number of moles x M_r = 0.0168 x 132 = **2.22 g**

10. The purity of commercially available sodium hydrogencarbonate was tested as follows. A 0.400g sample was dissolved in 100.0 cm³ of water and titrated against 0.200 M hydrochloric acid using methyl orange indicator.

$$NaHCO_3 + HCl \rightarrow NaCl + CO_2 + H_2$$

23.75 cm³ of acid was required for complete neutralisation.

Answer:

	$NaHCO_3$	+	HCl	→	$NaCl + CO_2 + H_2$
Reaction Coefficients	1		1		
Volume (cm³)	100		23.75		
No. of moles	**4.75 x 10⁻³**		23.75/1000 x 0.2 = 4.75 x 10⁻³		
Concentration (M)	1000/100 x 4.75 x 10⁻³ = 0.0475		0.2		
Mass (g)	0.0475 x 84 = 3.99 g				

a) How many moles of acid were used in the titration?

Answer:

Number of moles of HCl = volume (L) x concentration = 23.75/1000 x 0.2 = **4.75 x 10^{-3} moles**

b) Calculate the mass of sodium hydrogen carbonate titrated and hence the purity of the sample.

Answer:

Mass of 0.0475 moles of NaHCO$_3$ = number of moles x M$_r$ = 0.0475 x 84 = 3.99 g
% purity = mass of NaHCO$_3$/ mass of sample = 3.99/4.00 x 100 = **99.75%**

M Counting Moles .... Driving Test M

To Pass the Mole Driving Test and Throw Away Uour **M**-plates You Will Need to Achieve a Score of Greater than 70%. Each Question is Equally Weighted (5%).

Relative Atomic Masses: Hydrogen (H) = 1, Carbon (C) = 12, Nitrogen (N) = 14, Oxygen (O) = 16, Sodium (Na) = 23, Magnesium (Mg) = 24, Silicon (Si) = 28, Sulfur (S) = 32, Chlorine (Cl) = 35.5, Potassium (K) = 39, Titanium (Ti) = 48, Barium (Ba) = 137
Molar Volume: 22.4 L (22,4000 cm³) at STP (273.15 K and 1 atm)
Avogadro's Number: 6.02×10^{23} mol^{-1}

1. Calculate the mass, in grams, of a single atom of sodium-23.

Answer:

Mass of 6.02×10^{23} atoms of sodium = 23 g
Mass of one atom of Na = $23/6.02 \times 10^{23}$ = **3.82×10^{-23}**

2. An organic compound was found to contain 12.8% carbon, 2.13% hydrogen and 85.07% bromine. If the compound has a relative molecular mass of 188, determine its empirical formula and the molecular formula.

Answer:

	Carbon	Hydrogen	Bromine
% composition	12.8	2.13	85.07
A_r	12	1	80
% composition/A_r	12.8/12 = 1.06	2.13/1 = 2.13	85.07/80 =1.06
Ratio	1	2	1

Empirical Formula: CH_2Br
$M_r [CH_2Br]$ = 94
n = M_r [Molecular Formula]/M_r [Empirical formula] = 188/94 = 2

Molecular Formula = $(CH_2Br)n$ = $\mathbf{C_2H_4Br_2}$

3. What the mass of phosphorus is required to produce 200 g of phosphine (PH_3)?

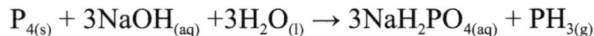

$$P_{4(s)} + 3NaOH_{(aq)} + 3H_2O_{(l)} \rightarrow 3NaH_2PO_{4(aq)} + PH_{3(g)}$$

Answer:

	$P_{4(s)}$	+	$3NaOH_{(aq)}$	+	$3H_2O_{(l)}$	→	$3NaH_2PO_{2(aq)}$	+	$PH_{3(g)}$
M_r	124		40		18		88		34
Mass Balance	124		120		54		264		
				298				298	
Reaction Coefficients	1		3		3		3		1
Mass (g)	5.88 x 124 = **729**								200
No. of moles	5.88								200/34 = 5.88

Number of moles in 200 g of PH_3 = mass/M_r = 200/34 = 5.88
According to the balanced chemical equation, 1 mole of PH_3 is produced from 1 mole of P_4
Therefore, 5.88 moles P_4 are required to produce 5.88 moles of PH_3
Mass of 5.88 moles P_4 = number of moles x M_r = 5.88 x 124 = **729 g**

4. A student heats 5.29g of $Sr(NO_3)_2$ and collects the gas.

$$2Sr(NO_3)_{2(s)} \rightarrow 2SrO_{(s)} + 4NO_{2(g)} + O_{2(g)}$$

Determine the volume of gas, at STP, obtained by the student. Molar mass of $Sr(NO_3)_2$ = 211.6 g mol^{-1}.

Answer:

	$2Sr(NO_3)_{2(s)}$	→	$2SrO_{(s)}$	+	$4NO_{2(g)}$	+	$O_{2(g)}$
M_r	211.6						
Reaction Coefficients	2		2		4		1
Mass	5.29						
No. of moles	5.29/211.6 = 0.025				0.05		0.0125

Number of moles in 5.29 g of $Sr(NO_3)_2$ = mass/M_r = 5.29/211.6 = 0.025

According to the balanced chemical reaction, 1 mole of $Sr(NO_3)_2$ generates 2.5 moles of gas

Thus 0.025 moles of $Sr(NO_3)_2$ will generate 2.5 x 0.025 = 0.0625 moles of gas

One mole of a gas at STP occupies a volume of 22400 cm^3 (V_m)

Therefore, 0.0625 moles will occupy a volume of 0.0625 x 22400 = **1400 cm^3**

5. A 12.41g sample of hydrated sodium thiosulfate, $Na_2S_2O_3$•$5H_2O$, was heated to remove the water of crystallization.

a) What is the relative formula mass of $Na_2S_2O_3$•$5H_2O$?

Answer:

Element	No. of atoms	A_r	Mass
Na	2	23	46
S	2	32	64
O	8	16	128
H	10	1	10
Total			139

M_r [$Na_2S_2O_3$•$5H_2O$] = **248**

b) Calculate the expected mass of anhydrous sodium thiosulfate that forms.

Answer:

M_r [$Na_2S_2O_3$•$5H_2O$] = 248
M_r [$Na_2S_2O_3$] = 158
Number of moles in 12.41 g of $Na_2S_2O_3$•$5H_2O$ = mass/M_r = 12.41/248 = 0.05
Number of moles of anhydrous $Na_2S_2O_3$ = 0.05
Mass = number of moles x M_r
Mass of 0.05 moles of $Na_2S_2O_3$ = 0.05 x 158 = **7.9 g**

6. The Kroll process is used to convert ore into titanium metal. Titanium chloride produced from the chlorination of ore, is reduced to titanium metal using magnesium under an inert atmosphere.

$$2Mg + TiCl_4 \rightarrow 2MgCl_2 + Ti$$

Calculate the maximum mass of titanium that could be produced from the addition of 3800 kg of titanium chloride to 1500 kg of magnesium.

Answer:

	2Mg	+	$TiCl_4$	→	$2MgCl_2$	+	Ti
M_r	24		190		95		48
Mass Balance	48		190		190		
		138				138	
Reaction Coefficients	2		1		2		1
Mass (g)	150000		380000				20,000 x 48 = **960000**
No. of moles	1500000/48 = 62,500		380000/190 = 20,000				20,000

Number of moles of $TiCl_4$ in 3800 kg (380000 g) = 380000/190 = 20,000
Note: Mg is in excess
According to the balanced chemical equation, 1 mole of $TiCl_4$ can produce one mole of Ti
Therefore, 20,000 moles of $TiCl_4$ can generate 20,000 moles of Ti
Mass = number of moles x M_r
Mass of 20,000 moles of Ti = 20,000 x 48 = 960,000 g = **960 kg**

7. A 1 mg sample of octane, C_8H_{18} was totally combusted in air.

a) How many moles are present in 1 mg of octane?

Answer:

1 mg = 0.001 g
Number of moles in 0.001 g of C_8H_{18} = mass / M_r = 0.001/114 = **8.77 x 10^{-6}**
b) Determine the number of moles and the volume of carbon dioxide generated.

Answer:

	C_8H_{18}	+	$12.5O_2$	→		$8CO_2$	+	$9H_2O$
M_r	114		32			44		18
Mass Balance	114	514	400			352	514	162

Table contd.....

	C_8H_{18}	+	$12.5O_2$	→	$8CO_2$	+	$9H_2O$
Reaction Coefficients	1		12.5		8		9
Mass (g)	1 x 10⁻³						
No. of moles	$0.001/114 = \textbf{8.77 x 10}^{-6}$				$8 \times 8.77 \times 10^{-6} = 7.02 \times 10^{-5}$		
Volume (cm³)					$7.02 \times 10^{-5} \times 22400 = \textbf{1.57}$		

Write a balanced chemical equation and construct a 'mole calculating frame' around it

According to the balanced chemical equation, 1 mole of C_8H_{18} will produce 8 moles of CO_2 on complete combustion

Therefore, the number of mole of CO_2 produced from the combustion of 8.77×10^{-6} moles of $C_8H_{18} = 8 \times 8.77 \times 10^{-6} = \textbf{7.02 x 10}^{-5}$

One mole of a gas occupies a volume of 22400 cm³

Therefore, 7.02×10^{-5} moles of CO_2 will occupy a volume of $7.02 \times 10^{-5} \times 22400 = \textbf{1.57}$ **cm³**

8.   Metal **M** forms a carbonate (M_2CO_3), which reacts with hydrochloric acid according to the following equation:

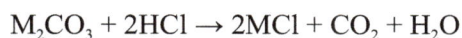

$$M_2CO_3 + 2HCl \rightarrow 2MCl + CO_2 + H_2O$$

A 0.394 g sample of M_2CO_3, was found to require the addition of 21.7 cm³ of a 0.263 M solution of hydrochloric acid (HCl) for complete reaction.

a)   Determine the number of moles of hydrochloric acid added.

Answer:

	M_2CO_3	+	$2HCl$	→	$2MCl$	+	CO_2	+	H_2O
Reaction Coefficients	1		2		2		1		1
Mass	0.394								
No. of moles			$21.7/1000 \times 0.263 = \textbf{5.7 x10}^{-3}$						
Concentration (M)			0.263						
Volume (cm³)			21.7						

Number of moles HCl in of 21.7 cm³ of a 0.263 M solution of hydrochloric acid = volume x concentration = $21.7/1000 \times 0.263 = \textbf{5.7 x10}^{-3}$

b) Determine the relative molecular mass of M_2CO_3 and hence the identity of M

Answer:

According to the balanced chemical equation, 1 mole of M_2CO_3 reacts with 2 moles of HCl

Therefore, the number of moles M_2CO_3 in 0.394 g = 0.5 x number of moles of HCl = 0.5 x 5.7 x10^{-3} = **2.85 x 10^{-3}**

Given that 2.85 x 10^{-3} moles M_2CO_3 has a mass of 0.394 g, then one mole (*i.e.* M_r) will have a mass of 1/2.85 x 10^{-3} x 0.394 = **138**

$M_r[CO_3]$= 60

Therefore 2 x A_r [M] = 138 – 60 = 78

A_r [M] = 78/2 = 39

Thus M = **potassium (K)**

9. An ammonia solution was reacted with sulfuric acid:

$$2NH_{3(aq)} + H_2SO_{4(aq)} \rightarrow (NH_4)_2SO_{4(aq)}$$

A 25.0 cm^3 aliquot of 1.24 M sulfuric acid required 30.8 cm^3 of this ammonia solution for complete reaction.

Calculate the concentration of the ammonia and the mass of ammonium sulfate present in the solution at the end of this titration.

Answer:

Number of moles in 25 cm^3 of 1.24 M H_2SO_4= volume (L) x concentration = 25/100 x 1.24 = 0.031

At the end point, number of moles of NH_3= 2 x number of moles of H_2SO_4 = 0.062

Thus the number of moles of NH_3 in 30.8 cm^3 = 0.062

Therefore, concentration of NH_3 = number of moles / volume (L) = 1000/30.8 x 0.062 = **2.01 M**

According to the balanced chemical equation, 2 moles of NH_3 produce 1 mole of $(NH_4)_2SO_4$

Therefore, 0.062 moles of NH_3 will produce 0,062 /2 = 0.031 moles of $(NH_4)_2SO_4$

Mass of 0.031 moles of $(NH_4)_2SO_4$ = number of moles x M_r

$M_r[(NH_4)_2SO_4] = 132$

Mass of $(NH_4)_2SO_4$ present at the end of the titration = 0.031 x 132 = **4.1 g**

10. Hydrogen is produced by the addition of hydrochloric acid to magnesium metal:

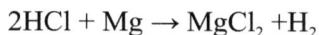

$$2HCl + Mg \rightarrow MgCl_2 + H_2$$

What mass of hydrogen is produced from the addition of 100 cm^3 of 5M hydrochloric acid of to an excess of magnesium?

Answer:

	2HCl	+	**Mg**	→	**MgC₁₂**	+		**H₂**
M_r	36.5		24		95			2
Mass Balance	73	97	24			97		
		97				97		
Reaction Coefficients	2		1		1			1
Volume (cm^3)	100							
Mass (g)								0.25 x 2 = **0.5**
No. of Moles	100/1000 x 5 = 0.5				0.25			0.5/2 = 0.25
Concentration (M)	5							

Number of moles in 100 cm^3 of 5M HCl = 100/1000 x 5 = 0.5 moles

According to the balanced chemical reaction, number of moles of H_2 = 0.5 x number of moles of HCl

Therefore, the maximum number of moles of H_2 generated from 0.5 moles of HCl = 0.5 x 0.5 = 0.25

Mass of 0.25 of H_2 = 0.25 x 2 = **0.5 g**

11. Ammonium nitrate is produced industrially by the reaction between ammonia and nitric acid:

$$NH_3 + HNO_3 \rightarrow NH_4NO_3$$

Calculate the volume of 2M nitric acid required to react with exactly 20.0 g of ammonia.

Answer:

	NH_3	+	HNO_3		$\rightarrow$	NH_4NO_3
M_r	17		63			80
Mass Balance		80				80
Reaction Coefficients	1		1			1
Mass (g)	20					
No. of moles	20/17 = 1.18		1.18			1.18
Volume (cm³)			1000/2 x 1.18 = 590			
Concentration (M)			2			

Number of moles in 20 g of NH_3 = 20/17 = 1.18

According to the balanced chemical equation, 1 mole of NH_3 reacts with 1 mole of HNO_3

Therefore, number of moles of HNO_3 that reacts exactly with 1.18 mole of NH_3 = 1.18

Volume of 2.00 M HNO_3 that contains 1.18 moles = number of moles/ concentration = 1.18/2 = **0.590 L (590 cm³)**

12.   Potassium chlorate, $KClO_3$, thermally decomposes according to the following equation:

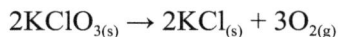

$$2KClO_{3(s)} \rightarrow 2KCl_{(s)} + 3O_{2(g)}$$

a) What mass of oxygen could be produced from the complete decomposition of 1.47 g of $KClO_3$?

Answer:

	$2KClO_3$	$\rightarrow$	$2KCl$	+	$3O_2$
M_r	122.5		74.5		32
Mass Balance	2 x 122.5 = 245		149		96
				245	
Reaction Coefficients	2		1		1
Mass (g)	1.47				0.018 x 32 = **0.576**
No. of Moles	1.47/122.5 = 0.012		0.012		1.5 x 0.12 = 0.018

Number of moles in 1.47 g of $KClO_3$ = 1.47/122.5 = 0.012
According to the balanced chemical equation, one mole of $KClO_3$ produces 1.5 moles of O_2
Therefore, 0.012 moles of $KClO_3$ could generate 1.5 x 0.12 = 0.018 moles of O_2
Mass of 0.018 moles of O_2 = 0.018 x 32 = **0.576 g**

b)   What mass of $KClO_3$ is required to generate 1.00 dm^3 of oxygen at STP?

Answer:

Note 1 L = 1 dm^3
Number of moles in 1L (1000 cm^3) of O_2 at STP = volume/V_m = 1000/22400 = 0.0446
According to the balanced chemical equation, 0.67 (*i.e.* 2/3) moles of $KClO_3$ are required
to produce 1 mole of O_2
Therefore, the number of moles of $KClO_3$ to produce 0.0446 moles O_2 = 0.0446 x 0.67 =
0.03 moles
Mass of 0.03 moles $KClO_3$ = number of moles x M_r = 0.03 x 122.5 = **3.65 g**

13.   25.0 cm^3 of 0.25 M sodium hydroxide required 22.5 cm^3 of a hydrochloric acid
solution for complete neutralisation. Calculate the concentration of the HCl solution.

Answer:

	NaOH	+	HCl	→	NaCl + H$_2$O
Reaction Coefficients	1		1		
Volume (cm³)	25		22.5		
No. of moles	25/1000 x 0.25 = 6.25 x 10⁻³		6.25 x 10⁻³		
Concentration (M)	0.25		1000/22.5 x 6.25 x 10⁻³ = **0.278**		

Number of moles in 25 cm^3 of 0.25 M NaOH = 25/1000 x 0.25 = 625 x 10^{-3}
At the end point, the number of moles of NaOH = number of moles of HCl
Therefore, the number of moles of HCl in 22.5 cm^3 = 6.25 x 10^{-3}
Concentration of HCl = 1000/22.5 x 6.25 x 10^{-3} = **0.278 M**

14.   Barium nitrate thermally decomposes as follows:

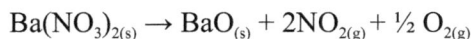

$$Ba(NO_3)_{2(s)} \rightarrow BaO_{(s)} + 2NO_{2(g)} + \tfrac{1}{2} O_{2(g)}$$

1. What is the total volume of gas, at STP, generated by the thermal decomposition of 5.00 g of barium nitrate?

Answer:

	$Ba(NO_3)_2$	→	BaO	+	$2NO_{2(g)}$	+	½ $O_{2(g)}$
M_r	261		153		46		16
Mass Balance	261				92		
						261	
Reaction Coefficients	1		1		2		0.5
Mass (g)	5						
No. of moles	5/261 = 0.01915				0.01915 x 2 = 0.038		0.01915 x 0.5 = 0.0096

Number of moles in 5 g of $Ba(NO_3)_2$ = mass/M_r = 5/261 = 0.01915
According to the balanced chemical equation, 1 mole of $Ba(NO_3)_2$ generates 2.5 moles of gas
Thus, 0.01915 moles of $Ba(NO_3)_2$ will generate 2.5 x 0.01915 = 0.048 moles of gas
Volume occupied at STP by 0.048 moles of gas = number of moles x V_m = 0.048 x 22400 = **1075 cm³**

b)   What volume of 1.20 M hydrochloric acid is required to neutralize the barium oxide produced by the thermal decomposition of 5.00 g of barium nitrate. Barium oxide reacts with hydrochloric acid as follows:

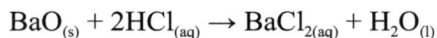

$$BaO_{(s)} + 2HCl_{(aq)} \rightarrow BaCl_{2(aq)} + H_2O_{(l)}$$

Answer:

	$BaO_{(s)}$	+	$2HCl_{(aq)}$	→	$BaCl_{2(aq)}$	+	$H_2O_{(l)}$
Volume (cm³)			1000/1.2 x 0.0383 = **31.9**				
No. of moles	0.01915		2 x 0.01915 = 0.0383				
Mass (g)							
Concentration (M)			1.21				

According to the balanced chemical equation, 1 mole of $Ba(NO_3)_2$ produces 1 mole of BaO

Thus, 0.01915 moles of $Ba(NO_3)_2$ will produce 0.01915 moles of BaO

1 mole of BaO requires 2 moles of HCl to be neutralized

Therefore, 2 x 0.01915 = 0.0383 moles of HCl are required to neutralize the BaO

Volume of 1.2 M required = number of moles/concentration = 0.0383/1.2 = **0.0319 L =**
31.9 cm³

15.   Silver nitrate thermally decomposes. 0.720 g of silver and 0.307 g of nitrogen dioxide was produced from heating a 1.133 g sample of silver nitrate in an open tube. The rest of the mass loss was due to oxygen. Using this data write the balanced chemical equation for the reaction.

Answer:

	$AgNO_3$	$\rightarrow$	Ag	+	NO_2	+	O_2
Reaction Coefficients	1		1		1		1
Mass	1.133		0.72		0.307		0.106
No. of Moles	0.0067		0.0067		0.0067		0.0033

Mass of oxygen released = 1.133 - 0.72 - 0.307 g = 0.106 g

Number of moles of $AgNO_3$ in 1.133 g = mass/M_r = 1.133/170 = 0.0067

Number of moles of Ag in 0.72 g = mass/M_r = 0.72/108 = 0.0067

Number of moles of NO_2 in 0.307 g = mass/M_r = 0.307/46 = 0.0067

Thus 0.0067 moles of $AgNO_3$ thermally decompose to produce 0.0067 moles Ag + 0.0067 moles NO_2 + 0.0033 moles O_2

Thus, 2 moles of $AgNO_3$ thermally degrade to produce 2 moles of Ag + 2 moles of NO_2 + 1 mole of O_2 **$2AgNO_{3/} \rightarrow 2Ag + 2NO_2 + O_2$**

PERIODIC TABLE

The Periodic Table of Elements

Key

relative atomic mass ⟶ $\boxed{\begin{array}{c}7 \\ \textbf{Li} \\ \text{lithium} \\ 3\end{array}}$

atomic number ⟶

1	2												3	4	5	6	7	0
1 **H** hydrogen 1																		4 **He** helium 2
7 **Li** lithium 3	9 **Be** beryllium 4												11 **B** boron 5	12 **C** carbon 6	14 **N** nitrogen 7	16 **O** oxygen 8	19 **F** fluorine 9	20 **Ne** neon 10
23 **Na** sodium 11	24 **Mg** magnesium 12												27 **Al** aluminium 13	28 **Si** silicon 14	31 **P** phosphorus 15	32 **S** sulphur 16	35.5 **Cl** chlorine 17	40 **Ar** argon 18
39 **K** potassium 19	40 **Ca** calcium 20	45 **Sc** scandium 21	48 **Ti** titanium 22	51 **V** vanadium 23	52 **Cr** chromium 24	55 **Mn** manganese 25	56 **Fe** iron 26	59 **Co** cobalt 27	59 **Ni** nickel 28	63.5 **Cu** copper 29	65 **Zn** zinc 30		70 **Ga** gallium 31	73 **Ge** germanium 32	75 **As** arsenic 33	79 **Se** selenium 34	80 **Br** bromine 35	84 **Kr** krypton 36
85 **Rb** rubidium 37	88 **Sr** strontium 38	89 **Y** yttrium 39	91 **Zr** zirconium 40	93 **Nb** niobium 41	96 **Mo** molybdenum 42	**Tc** technetium 43	101 **Ru** ruthenium 44	103 **Rh** rhodium 45	106 **Pd** palladium 46	108 **Ag** silver 47	112 **Cd** cadmium 48		115 **In** indium 49	119 **Sn** tin 50	122 **Sb** antimony 51	128 **Te** tellurium 52	127 **I** iodine 53	131 **Xe** xenon 54
133 **Cs** caesium 55	137 **Ba** barium 56	139 **La** lanthanum 57 ❶	178 **Hf** hafnium 72	181 **Ta** tantalum 73	184 **W** tungsten 74	186 **Re** rhenium 75	190 **Os** osmium 76	192 **Ir** iridium 77	195 **Pt** platinum 78	197 **Au** gold 79	201 **Hg** mercury 80		204 **Tl** thallium 81	207 **Pb** lead 82	209 **Bi** bismuth 83	**Po** polonium 84	**At** astatine 85	**Rn** radon 86
Fr francium 87	226 **Ra** radium 88	227 **Ac** actinium 89 ❷																

❶ 58 – 71 Lanthanum series

❷ 90 – 103 Actinium series

140 **Ce** cerium 58	141 **Pr** praseodymium 59	144 **Nd** neodymium 60	**Pm** promethium 61	150 **Sm** samarium 62	152 **Eu** europium 63	157 **Gd** gadolinium 64	159 **Tb** terbium 65	162 **Dy** dysprosium 66	165 **Ho** holmium 67	167 **Er** erbium 68	169 **Tm** thulium 69	173 **Yb** ytterbium 70	175 **Lu** lutetium 71
232 **Th** thorium 90	**Pa** protoactinium 91	238 **U** uranium 92	**Np** neptunium 93	**Pu** plutonium 94	**Am** americium 95	**Cm** curium 96	**Bk** berkelium 97	**Cf** californium 98	**Es** einsteinium 99	**Fm** fermium 100	**Md** mendelevium 101	**No** nobelium 102	**Lr** lawrencium 103

All rights reserved-© 2016 Bentham Science Publishers

License

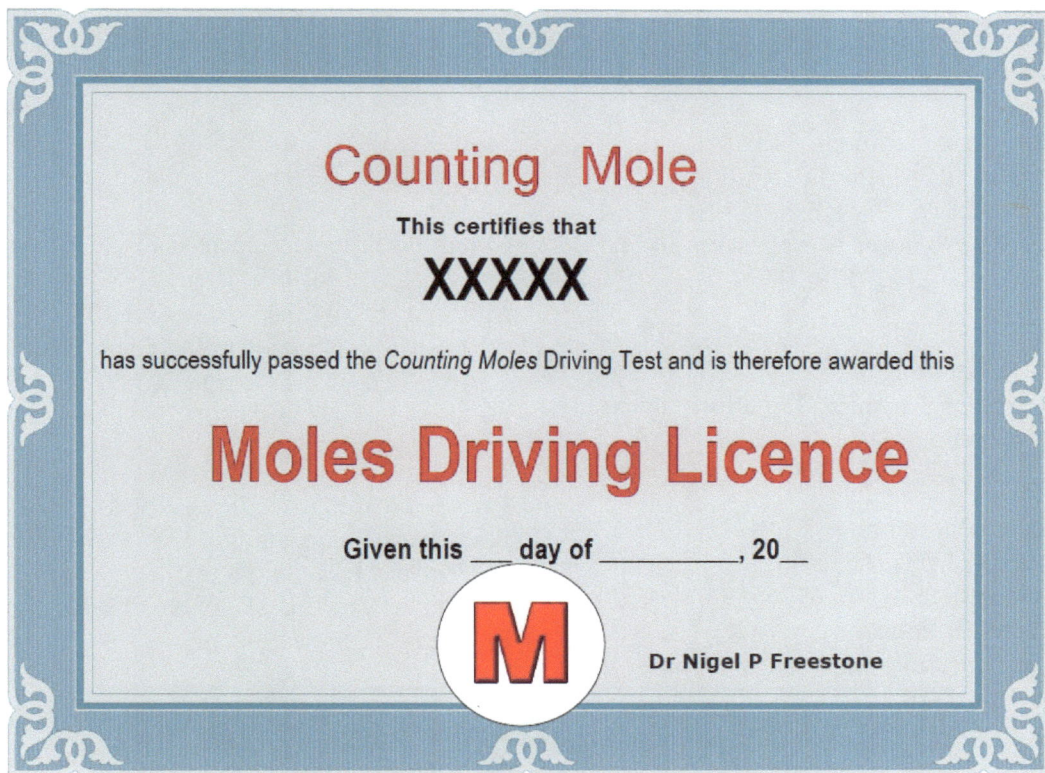

All rights reserved-© 2016 Bentham Science Publishers

SUBJECT INDEX

Nigel P. Freestone
All rights reserved-© 2016 Bentham Science Publishers

www.ingramcontent.com/pod-product-compliance
Lightning Source LLC
Chambersburg PA
CBHW050840220326
41598CB00006B/411

* 9 7 8 1 6 8 1 0 8 1 0 1 4 *